SUSTAINABILITY

T0332366

Analytics and Control Series

Series Editors:
Adedeji B. Badiru, Air Force
Institute of Technology, Dayton, Ohio, USA

Decisions in business, industry, government, and the military are predicated on performing data analytics to generate effective and relevant decisions, which will inform appropriate control actions. The purpose of the focus series is to generate a collection of short form books focused on analytic tools and techniques for decision making and related control actions.

Mechanics of Project Management
Nuts and Bolts of Project Execution
Adedeji B. Badiru, S. Abidemi Badiru, and I. Adetokunboh Badiru

The Story of Industrial Engineering
The Rise from Shop-Floor Management to Modern Digital Engineering
Adedeji B. Badiru

Innovation
A Systems Approach
Adedeji B. Badiru

Project Management Essentials
Analytics for Control
Adedeji B. Badiru

Sustainability
Systems Engineering Approach to the Global Grand Challenge
Adedeji B. Badiru and Tina Agustiady

For more information on this series, please visit:https://www.routledge.com/Analytics-and-Control/book-series/CRCAC

SUSTAINABILITY

A Systems Engineering Approach to the Global Grand Challenge

Adedeji B. Badiru and Tina Agustiady

CRC Press
Taylor & Francis Group
Boca Raton London New York

CRC Press is an imprint of the
Taylor & Francis Group, an **informa** business

First edition published 2021
by CRC Press
6000 Broken Sound Parkway NW, Suite 300, Boca Raton, FL 33487-2742

and by CRC Press
2 Park Square, Milton Park, Abingdon, Oxon, OX14 4RN

© 2021 Taylor & Francis Group, LLC

CRC Press is an imprint of Taylor & Francis Group, LLC

The right of Adedeji B. Badiru and Tina Agustiady to be identified as authors of this work has been asserted by them in accordance with sections 77 and 78 of the Copyright, Designs and Patents Act 1988.

Library of Congress Cataloging-in-Publication Data
Names: Badiru, Adedeji Bodunde, 1952- author. | Agustiady, Tina, author.
Title: Sustainability : a systems engineering approach to the global grand
challenge / Adedeji B. Badiru and Tina Agustiady.
Description: Boca Raton : CRC Press, 2021. |
Series: Analytics and control | Includes bibliographical references and index.
Identifiers: LCCN 2020046767 (print) | LCCN 2020046768 (ebook) | ISBN 9780367431211
(hardback) | ISBN 9781003005025 (ebook)
Subjects: LCSH: Sustainable engineering. | Sustainable development.
Classification: LCC TA163 .B33 2021 (print) | LCC TA163 (ebook) | DDC 658.4/013--dc23
LC record available at https://lccn.loc.gov/2020046767
LC ebook record available at https://lccn.loc.gov/2020046768

ISBN: 978-0-367-43121-1 (hbk)
ISBN: 978-0-367-74780-0 (pbk)
ISBN: 978-1-003-00502-5 (ebk)

Typeset in Times
by MPS Limited, Dehradun

Dedicated to our children, who shall carry on the pursuit of sustainability

Contents

Contents

Preface

Sustainability has been a subject of worldwide interest and concern for many years. In spite of the various approaches that have been advanced more recently, much remains to be done. Initially, we were planning to follow the path of previous writings on sustainability. But then, the COVID-19 pandemic struck and it brought on new angles and perspectives on the pursuit of sustainability. No matter what sustainability entails in concept, ideas, goals, expectations, and objectives, ultimately, it will be the business and industry enterprises that will carry the burden of supporting sustainability through business operations. By inference and extension, the personal-level sustainability commitment can follow. Sustainability often automatically connotes environmental sustainability. But sustainability actually goes beyond environmental issues. In many cases, the issues that cause sustainability concerns have their roots in the practices that exist in the business and industry environments. Consequently, our focus for this book is to leverage business survival, through systems thinking, as the basis for addressing sustainability. Everything we do as humans, whether at home, in business, in industry, in academia, in the military, or in the government, eventually impacts the environment in one form or another. Thus, the methodology of this book is to address the sources of sustainability from a business survival perspective. We use two complementary platforms as the premise for the contents of the book. The first platform is the 14 Grand Challenges for Engineering published by the US National Academy of Engineering (NAE) in 2008. The second platform is the use of a systems framework to address the diverse global elements affecting sustainability. Specifically, the Design, Evaluate, Justify, and Integrate (DEJI) systems model® is adopted for this book's structure.

The NAE's 14 grand challenges present opportunities and challenges for business and industry. The topic of sustainability permeates the NAE grand challenges. Our belief is that companies can make contributions toward addressing the grand challenges by practicing better business protocols and strategies. Benefits are always needed in companies in order to be successful and profitable. While continuous improvement techniques are desired, the implementation of the techniques become difficult and challenging to sustain, if we do not view things from a systems perspective. The proper tools form the key to making decisions and having successful results. Sustainability requires project management and accountability along with matrices to understand the order of events. In order to sustain results, simply relying on an individual or one particular dataset or tool is not enough. True statistical techniques need to be implemented to help make each industry the best in what they do, with a focus on sustainability. Lean and Six Sigma are important tools that have been proven to be results-oriented. Thus, they can help achieve results in sustainability programs. The emerging knowledge economy and the move toward digital engineering need a systems approach that is aligned with the 14 grand challenges. The chapters of this book present tools and techniques for that purpose.

Adedeji B. Badiru and Tina Agustiady

Acknowledgments

We thank our families for standing by us patiently throughout the prolonged journey of preparing this manuscript.

Authors

Adedeji B. Badiru is the Dean and Senior Academic Officer for the Graduate School of Engineering and Management at the Air Force Institute of Technology (AFIT) located at Wright-Patterson AFB, Ohio. He was previously Professor and Head of Systems Engineering and Management at AFIT, Professor and Department Head of Industrial & Information Engineering at the University of Tennessee in Knoxville, and Professor of Industrial Engineering and Dean of University College at the University of Oklahoma, Norman. He is a registered professional engineer, a certified Project Management Professional, a Fellow of the Institute of Industrial Engineers, and a Fellow of the Nigerian Academy of Engineering. He holds a BS degree in Industrial Engineering, an MS degree in Mathematics, an MS degree in Industrial Engineering from Tennessee Technological University, and a PhD degree in Industrial Engineering from the University of Central Florida.

His areas of interest include mathematical modeling, project modeling and analysis, economic analysis, systems engineering, and efficiency/productivity analysis and improvement. He is the author of over 35 books, 38 book chapters, 88 technical journal articles, and 220 conference proceedings and presentations.

Often venerated by his colleagues as "a writing machine," Badiru has also published 35 magazine articles and 20 editorials and periodicals. He is a member of several professional associations and scholastic honor societies. He is the Series Editor for the Taylor and Francis book series on Systems Innovation and Focus series on Analytics and Control. Professor Badiru was the 2020 winner of Taylor & Francis Lifetime Achievement Award.

Tina Agustiady is a certified Six Sigma Master Black Belt and Continuous Improvement Leader.

Her many activities and responsibilities include serving as the past Lean division president and current technical vice president of Institute of Industrial and Systems Engineers (IISE). She has served as a board director and chairman for the IISE annual conferences and Lean Six Sigma conferences.

Agustiady is an instructor who trains and certifies students for Lean, Six Sigma, Innovation, Design of Experiments and Business Process

Management for IISE, Lean Sigma Corporation, Six Sigma Digest, and Villanova/Bisk University. She is a Subject Matter Expert, Professor and Course Designer for professional certification programs, including undergraduate and graduate degrees at Villanova University and Bisk Education.

Her accomplishments as a writer and author include serving as an editor for the *International Journal of Six Sigma and Competitive Advantage*. She has co-authored *Statistical Techniques for Project Control*, *Sustainability: Utilizing Lean Six Sigma Techniques*, and *Total Productive Maintenance: Strategies and Implementation Guide*. She also authored *Communication for Continuous Improvement Projects* and her recently published book *Design for Six Sigma: A Practical Approach through Innovation*. She is a CRC Press/ Taylor and Francis series editor for Continuous Improvement and Lean Innovation and Management.

She was honored to be a Feigenbaum medalist for 2016 and a Crosby medalist for 2018.

Systems Framework for Global Challenges

1

"To be prepared is to be confident in sustainability."

— Adedeji Badiru

INTRODUCTION: SYSTEMS VIEW OF THE WORLD

Buzz Lightyear, a fictional character in the *Toy Story* movies, said, "To infinity…and beyond!" We say, "Sustainability to infinity and beyond." It is a systems world nowadays. COVID-19 has revealed the world's fragility and we must take action from a world-systems perspective. The pursuit of sustainability is a collective responsibility, which we must work together to accomplish. Henry Ford, the auto industry pioneer, said it best with the quote, "If everyone is moving forward together, then success takes care of itself."

More concerted global-based efforts are needed to right the ship of sustainability. The world is more intertwined now, more than ever before. Fast travels and digital connectivity have created a faster influence on the environment globally. The emergence of COVID-19 pandemic in early 2020 further exacerbated the global influence in terms of health, commerce, business, industry, government, education, law enforcement, and the military. In essence, this time is unlike any other time in the history of the world. Everyone is affected in every nook and corner of the world. To this extent, sustainability, in whichever way we define it, is threatened. This creates the imperative that we must use new and hybrid approaches to address the issues and concerns of sustainability. In this book, we take the approach of enmeshing a systems model with the global platform provided by the 14 Grand Challenges for Engineering published by the US National Academy of Engineering (NAE) in 2008. This hybrid approach is more robust as it is applied to the business and industry performance environment, which can serve as the anchor of sustainability. The global environmental problems of climate change and sustainability affect everyone on our planet. Hence, we must all combine our efforts together, as a system, to evolve collective and lasting solutions.

Figure 1.1 illustrates the global reach of operational excellence. Business, industry, government, academia, and the military are all desirous of sustainable operational excellence.

FIRST PRINCIPLES OF SUSTAINABILITY

We must do something for the sustainability to take hold. Doing nothing means achieving nothing. Every little effort incrementally adds to the total result. This is conveyed by the simple mathematical thinking expressed below, without the rigor of any mathematical proof: 0 to the power of 365=0, 1 to the power of 365=1, and 1.01 to the power of 365=37.7. So, if we can do something small toward sustainability every day of the year (365 days), we will get somewhere along the path of achieving our goal. As a first principle, sustainability must be communicated explicitly to everyone with a high level of clarity and understandability. In essence, sustainability is more than just environmental concerns. A systems view of sustainability encompasses examining the various human activities that, ultimately, impact the environment potentially in adverse ways. By going to the first principles of sustainability, we can tackle the elemental sources of adverse impacts on the environment. For example, the first principles of engineering are foundational propositions and assumptions that lead us to the very root of theories. When first principles are applied, the belief is that, if we understand the pieces of the system, we understand the system. Therein lies the premise of this book in advocating a systems view of sustainability. Although some of the contents in this book might have been seen or read in different contexts before, reinforcement is a good thing for achieving a global success with sustainability. Repeatability is one of the expectations of a systems approach.

One global effort that is worth mentioning here is the annual Tang Prize in Sustainable Development. The prize was established in 2012 by

FIGURE 1.1   In–Out Interconnectivity of the World System for Sustainability. Source: © Can Stock Photo/Lemony.

Dr. Samuel Yen-Liang Yin, a Taiwanese billionaire businessman and philanthropist. The prize is for the purpose of encouraging global citizens and organizations to tackle the challenges unique to the 21st century so that we can ensure the sustainable development of mankind and nature. The 2020 prize was awarded to Dr. Jane Goodall in recognition of her groundbreaking discovery in primatology that redefines human–animal relationship and for her lifelong unparalleled dedication to the conservation of Earth environment. From a global systems perspective, we need more individual and organizational efforts like this in order for us to achieve the ideals of sustainability.

Swinging across the oceans from Taiwan, we can cite one community example in the City of Dayton, Ohio, USA. The city identified efficiency and resiliency as the keys to sustainability. In 2020, the Dayton City Commission approved Dayton's first comprehensive sustainability strategy, within which 115 projects or initiatives would be pursued for positioning Dayton to actualize the following goals and objectives:

1. Be efficient in energy and resource consumption
2. Transition to renewable energy as quickly as feasible and affordable
3. Foster diverse transportation options
4. Encourage the right type of recycling
5. Ensure wise use of land resources, urban agriculture, and building structures

An initial tracking period of five years was endorsed for Dayton's Sustainability Office, which periodically reports progress via its website daytonohio.gov/sustainability under a banner of "Strategy for a Sustainable Dayton." No doubt, many other cities are undertaking similar efforts. With all these local efforts, however small they may be, coming together as a system, the world can move forward rapidly in sustainability. We are particularly enthused by the city's focus on efficiency and resiliency because efficiency and resiliency, from a systems perspective, form a core of the tools and techniques of industrial engineering, which we are advocating as the theme of this book.

SYSTEM CONTEXT OF SUSTAINABILITY

A system is a collection of interrelated elements working together synergistically to achieve a set of objectives. Any project is essentially a collection of interrelated activities, people, tools, resources, processes, and other assets brought together in the pursuit of a common goal. The goal may be in terms of generating a physical product, providing a service, or achieving a specific result. This makes it possible to view any project as a system that is amenable to all the classical and modern concepts of systems management. A systems view of the world makes sustainability to work better and projects more likely to succeed (Agustiady and Badiru 2013; Badiru 2019). A systems view provides a disciplined process for the design, development, and execution of complex sustainability initiatives in business, industry, education, and government.

A major advantage of a systems approach to sustainability is the win–win benefit for everyone. A systems view also allows full involvement of all stakeholders of sustainability.

Systems engineering is the application of engineering to solutions of a multifaceted problem through a systematic collection and integration of parts of the problem with respect to the lifecycle of the problem. It is the branch of engineering concerned with the development, implementation, and use of large or complex systems. It focuses on specific goals of a system, considering the specifications, prevailing constraints, expected services, possible behaviors, and structure of the system. It also involves a consideration of the activities required to assure that the system's performance matches the stated goals. Systems engineering addresses the integration of tools, people, and processes required to achieve a cost-effective and timely operation of the system. We are all stakeholders of sustainability. As such, we are subcomponents of a large societal system. Systems engineering is concerned with the big picture of the challenges that we face. This involves how a system functions or behaves overall, how it interfaces with its users, how it responds to its environment, how it interacts with other systems, and how it regulates itself. Sustainability must be approached both from a technical point of view and from a management viewpoint. The management discipline organizes and allocates efforts and resources across the broad spectrum of the system, including initiating communication, facilitating collaboration, defining system requirements, planning work flows, and deploying technology for targeted needs. A systems engineering framework for sustainability has the following elements:

- Focus on the end goal
- Involve all stakeholders
- Define the sustainability issue of interest
- Break down the problem into manageable work packages
- Connect the interface points between project requirements and project design
- Define the work environment to be conducive to the needs of participants
- Evaluate the systems structure
- Justify every major stage of the sustainability project
- Integrate sustainability into the core functions existing in the organization

MULTIDIMENSIONALITY OF SUSTAINABILITY

Different experts, researchers, and practitioners have differing definitions of sustainability. The lack of a consistent view is probably the reason that many sustainability programs have not taken root as expected. The following systems-based definitions are essential for the purpose of the theme of this book.

Sustainability – The human preservation of the environment, whether economically or socially through responsibility, management of resources, and maintenance of physical infrastructure.

Recycling – The reprocessing of materials already used into new materials or products to prevent waste and protect the environment by reducing energy, air pollution, water pollution, and emissions.

Reducing Waste – Reducing the amount of unwanted materials technologically and socially to economically benefit the environment.

Reuse – To use again in a different circumstance after processing.

Energy – The capacity of a physical system to perform work through heat, kinetics, mechanical systems, light, or electrical means.

Sustainable Energy – The condition of energy that meets the needs of the present without compromising the ability of future generations to meet their needs.

Environment – The setting, surroundings, or conditions in which living objects operate.

Ecosystem – Interacting organisms and their physical means of living in a biological community.

Humanity – Human nature and civilization.

Global – Relation to the world, Earth, or planet.

Global Warming – The gradual increase in temperature of the Earth and oceans predicted to be from pollution and the inconsideration of the environment.

Greenhouse – Solar radiation entrapment caused by atmospheric gases caused by pollution, which allows sunlight to pass through and be remitted as heat radiation back from the Earth's surface.

Quality Management – The key for a system to ensure the end goal is met through a desired level of excellence to be competitive in the business.

Total Quality Management – Continuous improvement in products and processes by increasing the quality and reducing the defects through management methodologies.

Systems – A network of things interacting together.

Biodiversity – The natural variation among living organisms in particular ecosystems.

Climate – Temperature, precipitation, and wind characteristics and conditions.

Natural Resources – Materials or matter in nature such as minerals, fresh water, forests, or abundant land that can be used for economic benefits.

Forecasting – The prediction or estimation of future events performed normally due to trending.

Agriculture – Farming and manipulation of soils to harvest and grow crops while raising livestock.

Human Impact – The impacts from human beings that affect biophysical environments, biodiversity, and any other environments.

Public Health – The science and art of preventing diseases while prolonging life.

Project Management – Planning, managing, monitoring, and controlling projects utilizing feedback and knowledge of tools and techniques.

Ethics – A system of moral values that makes one perform the right conduct.

Education – Education or training performed through knowledge and studying.

Policies – Plans or courses of actions.

Physics – The science of matter and energy and the interaction of the two.

Cooperation – The process of two or more beings working together toward the same goal.

Coordination – Organization of different elements so that there is cooperation for effective processes.

Planetary – Relating to the Earth as a planet.

Systems – Detailed methods or procedures established to transmit out a specific activity or to perform a responsibility.

Theory – A set of assumptions based on accepted facts that provide rational explanations of cause and effect relationships among groups.

Community Service – Voluntary work intended to assist others in a particular area.

Enterprise – A business or company with resources.

Biosphere – The actual regions where living organisms occupy or reside.

Biophysical – The science dealing with the application of physics of biological processes.

Sustainable Development – The concept of needs mostly from the idea of limitations imposed by society and technology to prevent the environment from meeting present and future needs.

Society – The collective group of people living together in a particular region with the same customs, laws, or organizations.

Economy – The prosperity, possessions, and resources of a country or region in terms of production and consumption of goods and services.

Energy Efficient Coding – Codes that set minimum requirements for energy-efficient design and construction for new and renovated buildings that impact energy usage.

Fenestration – The arrangement of windows and doors in a building to help operate with lower heating and cooling losses.

PERSONAL ACCOUNTABILITY FOR SUSTAINABILITY

Demonstrating personal accountability for sustainability is essential for a global success. Sustainability is not just for the environment. Although environmental concern is what immediately comes to mind whenever the word "sustainability" is mentioned, there are many languages (i.e., modes) of sustainability, depending on whatever perspective is under consideration. The context determines the interpretation. Each point of reference determines how we, as individuals or groups, respond to the need for sustainability. Pursuits of green building, green engineering, clean water, climate research, energy conservation, eco-manufacturing, clean product design, lean production, and so on remind us of the foundational importance of sustainability in all we do.

Commitment to sustainability is in vogue these days, be it in the corporate world or personal pursuits. But what exactly is sustainability? Definitions of the word contain verbs, nouns, and adjectives such as "green," "clean," "maintain," "retain," "stability," "ecological balance," "natural resources," and "environment." The definition of sustainability implies the ability to sustain (and maintain) a process or object at a desirable level of utility. The concept of sustainability applies to all aspects of functional and operational requirements, embracing both technical and managerial needs. Sustainability requires methodological, scientific, and analytical rigor to make it effective for managing human activities and resources.

In the aforementioned context, sustainability is nothing more than prudent resource utilization. The profession of industrial engineering is uniquely positioned to facilitate sustainability, especially as it relates to the environment, technical resources, management processes, human interfaces, product development, and facility utilization. Industrial engineers have creative and simple solutions to complex problems. Sustainability is a complex undertaking that warrants the attention and involvement of industrial engineers. A good example of the practice of sustainability is how a marathon runner strategically expends stored energy to cover a long-distance race. Burning up energy too soon means that the marathon race will not be completed. Erratic expenditure of energy would prevent the body from reaching its peak performance during the race. Steady-state execution is a foundation for achieving sustainability in all undertakings where the decline of an asset is a concern. An example is an analysis of how much water or energy it takes to raise cattle for human consumption. Do we ever wonder about this as we delve into our favorite steaks? Probably not. Yet this is, indeed, an issue of sustainability.

For this reason, we present the following succinct definition of the role of industrial engineers in sustainability:

> **Industrial engineers make systems function better together with less waste, better quality, fewer resources, and on target with goals, objectives, and requirements.**

RESOURCE CONSCIOUSNESS IN SUSTAINABILITY

The often-heard debate about what constitutes sustainability can be alleviated if we adopt the context of "resource consciousness," which, in simple terms, conveys the pursuit of conservation in managing our resources. All the resources that support our objectives and goals are amenable to sustainability efforts. For example, the expansion of a manufacturing plant should consider sustainability, not only in terms of increased energy consumption but also in terms of market sustainability, intellectual property sustainability, manpower sustainability, product sustainability, and so on. The limited resource may be spread too thin to cover the increased requirements for a larger production facility. Even a local community center should consider sustainability when contemplating expansion projects just as the local government should consider tax base sustainability when embarking on new programs. The mortgage practices that led to the housing industry bust in the United States were due to financial expectations that were not sustainable. If we put this in the context of energy consumption, it is seen that buying a bigger house implies a higher level of energy consumption, which ultimately defeats the goal of environmental sustainability. Similarly, a sports league that chooses to expand haphazardly will eventually face nonsustainability dilemma. Every decision ties back to the conservation of some resource (whether a natural resource or a manufactured resource), which links directly to the conventional understanding of sustainability.

FOCI OF SUSTAINABILITY

There are several moving parts in sustainability. Only a systems view can ensure that all components are factored into the overall pursuit of sustainability.

A systems view of the world allows an integrated design, analysis, and execution of sustainability projects. It would not work to have one segment of the world embarking on sustainability efforts while another segment embraces practices that impede overall achievement of sustainability. In the context of production for the global market, whether a process is repeatable or not, in a statistical sense, is an issue of sustainability. A systems-based framework allows us to plan for prudent utilization of scarce resources across all operations. Some specific areas of sustainability include the following:

- Environmental sustainability
- Operational sustainability
- Energy sustainability
- Health and welfare sustainability
- Safety and security sustainability
- Market sustainability
- Financial sustainability
- Economic sustainability
- Health sustainability
- Family sustainability
- Social sustainability

The long list of possible areas means that sustainability goes beyond environmental concerns. Every human endeavor should be planned and managed with a view toward sustainability.

VALUE SUSTAINABILITY

Sustainability imparts value on any organizational process and product. Even though the initial investment and commitment to sustainability might appear discouraging, it is a fact that sustainability can reduce long-term cost, increase productivity, and promote achievement of global standards. Sample questions for value sustainability are provided as follows:

- What is the organizational mission in relation to the desired value stream?
- Are personnel aware of where value resides in the organization?
- Will value assignment be on team, individual, or organizational basis?

- Is the work process stable enough to support the acquisition of value?
- Can value be sustained?

LEVERAGING THE HIERARCHY OF NEEDS

Sustainability is often tied to the basic needs of humans. The psychology theory of "Hierarchy of Needs" postulated by Abraham Maslow in his 1943 paper, "A Theory of Human Motivation," still governs how we respond along the dimensions of sustainability, particularly where group dynamics and organizational needs are involved. An environmentally induced disparity in the hierarchy of needs implies that we may not be able to fulfill the personal and organizational responsibilities along the spectrum of sustainability. In a diverse workforce, the specific levels and structure of needs may be altered from the typical steps suggested by Maslow's Hierarchy of Needs. This calls for evaluating the needs from a multidimensional perspective. For example, a three-dimensional view of the hierarchy of needs can be used to coordinate personal needs with organizational needs with the objective of facilitating sustainability. People's hierarchy of basic needs will often dictate how they respond to calls for sustainability initiatives. Maslow's hierarchy of needs consists of the following five stages:

1. **Physiological needs:** These are the needs for the basic necessities of life, such as food, water, housing, and clothing (i.e., survival needs). This is the level where access to money is most critical. *Sustainability applies here.*
2. **Safety needs:** These are the needs for security, stability, and freedom from physical harm (i.e., desire for a safe environment). *Sustainability applies here.*
3. **Social needs:** These are the needs for social approval, friends, love, affection, and association (i.e., desire to belong). For example, social belonging may bring about better economic outlook that may enable each individual to be in a better position to meet his or her social needs. *Sustainability applies here.*
4. **Esteem needs:** These are the needs for accomplishment, respect, recognition, attention, and appreciation (i.e., desire to be known). *Sustainability applies here.*
5. **Self-actualization needs:** These are the needs for self-fulfillment and self-improvement (i.e., desire to arrive). This represents the

stage of opportunity to grow professionally and be in a position to selflessly help others. *Sustainability applies here.*

Ultimately, the need for and commitment to sustainability boil down to each person's perception based on his or her location on the hierarchy of needs and level of awareness of sustainability. How do we explain to a hungry poor family in an economically depressed part of the world the need to conserve forestry? Or, how do we dissuade an old-fashioned professor from the practice of making volumes of hardcopy handouts instead of using electronic distribution? Cutting down on printed materials is an issue of advancing sustainability. In each wasteful eye, "the *need* erroneously justifies the *means*" (author's own variation of the common phrase). This runs counter to the principle of sustainability. We can expand the hierarchy of needs to generate a multidimensional view that incorporates organizational hierarchy of needs. The location of each organization along its hierarchy of needs will determine how the organization perceives and embraces sustainability programs. Likewise, the hierarchy position of each individual will determine how he or she practices commitment to sustainability.

In an economically underserved culture, most workers will be at the basic level of physiological needs, and there may be constraints on moving from one level to the next higher level. This fact has an implication on how human interfaces impinge upon sustainability practices. In terms of organizational hierarchy of needs, the levels are characterized as follows:

Level 1 of organizational needs: This is the organizational need for basic essentials of economic vitality to support the provision of value for stockholders and employees. Can the organization fund projects from cash reserves? *Sustainability applies here.*

Level 2 of organizational needs: This is a need for organizational defense. Can the organization feel safe from external attack? Can the organization protect itself from cyberattacks or brutal takeover attempts? *Sustainability applies here.*

Level 3 of organizational needs: This is the need for an organization to belong to some market alliances. Can the organization be invited to join trade groups? Does the organization have a presence on some world stage? *Sustainability applies here.*

Level 4 of organizational needs: This is the level of having market respect and credibility. Is the organization esteemed in some aspect of market, economic, or technology movement? What positive thing is the organization known for? *Sustainability applies here.*

Level 5 of organizational needs: This is the level of being classified as a "Power" in the industry of reference. Does the nation have a recognized niche in the market? *Sustainability applies here.*

Obviously, where the organization stands in its hierarchy of sustainability goals will determine how it influences its employees (as individuals) to embrace, support, and practice sustainability. How each individual responds to organizational requirements depends on that individual's own level in the hierarchy of needs. We must all recognize the factors that influence sustainability in our strategic planning programs. In order for an organization to succeed, sustainability must be expressed explicitly as a goal across organizational functions.

SUSTAINABILITY MATRIX

The coupling of technical assets and managerial tools are essential for realizing sustainability. This section presents an example of the *sustainability matrix* introduced by Badiru (2010). The matrix is a simple tool for organizing the relevant factors associated with sustainability. It overlays sustainability awareness factors, technical assets, and managerial tools. The sample elements illustrate the nature and span of factors associated with sustainability projects. Each organization must assess its own environment and include relevant factors and issues within the context of prevailing sustainability programs. Without a rigorous analytical framework, sustainability will just be in talks rather than deeds. One viable strategy is to build collaborative science, technology, engineering, andm alliances for sustainability projects. The analytical framework of systems engineering provides a tool for this purpose from an interdisciplinary perspective. With this, environmental systems, industrial systems, and societal systems can be sustainably tied together to provide win–win benefits for all. An effective collaborative structure would include researchers and practitioners from a wide variety of disciplines (civil and environmental engineering, industrial engineering, mechanical engineering, public health, business, etc.).

Project sustainability is as much a need as the traditional components of project management spanning planning, organizing, scheduling, and control. Proactive pursuit of the best practices of sustainability can pave the way for project success on a global scale. In addition to people, technology, and process issues, there are project implementation issues. In terms of performance, if we need a better policy, we can develop it. If we need technological advancement, we have capabilities to achieve it. The items that are often beyond reach relate to project lifecycle management issues. Project sustainability implies that sustainability exists in all factors related to the project. Thus, we should always focus on project sustainability.

The vertical dimension of the matrix consists of the following *technical factors*:

- Physical infrastructure
- Work design
- Analytical modeling
- Scientific limitation
- Technology constraints

The horizontal dimension of the matrix consists of the following *managerial environmental factors*:

- Organizational behavior
- Personnel culture
- Resource base
- Market influence
- Share capital

The cells within the matrix consist of a variety of attributes, factors, and indicators, including the following: Communication Modes, Cooperation Incentives, Coordination Techniques, Building Performance, Energy Economics, Technical Acquisitions, Work Measurement, Project Design, Financial Implications, Project Control, Resource Combinations, Qualitative Risk, Engineering Analysis, Value Assessment, Forecast Models, Fuel Efficiency, Technical Workforce, Contingency Planning, Contract administration, Green Purchases, Energy Conservation, Training Programs, Quantitative Risk, Public Acceptance, and Technology Risks.

Think "sustainability" in all you do and you are bound to reap the rewards of better resource utilization, operational efficiency, and process effectiveness. Both management and technical issues must be considered in the pursuit of sustainability. People issues must be placed at the nexus of all the considerations of sustainability. Otherwise, sustainability itself cannot be sustained. Many organizations are adept at implementing rapid improvement events (RIEs). This chapter recommends a move from mere RIEs to sustainable improvement events.

SOCIAL CHANGE FOR SUSTAINABILITY

Change is the root of advancement. Sustainability requires change. Our society must be prepared for change in order to achieve sustainability. Efforts that support sustainability must be instituted into every aspect of everything that the society does. If society is better prepared for change, then positive changes can be achieved. The "pain but no gain" aspects of sustainability can be avoided if proper preparations have been made for societal changes. Sustainability requires an increasingly larger domestic market to preserve precious limited natural resources. The social systems that make up such markets must be carefully coordinated. The socioeconomic impact on sustainability cannot be overlooked.

Social changes are necessary to support sustainability efforts. Social discipline and dedication must be instilled in the society to make sustainability changes possible. The roles of the members of a society in terms of being responsible consumers and producers of consumer products must be outlined. People must be convinced of the importance of the contribution of each individual whether that individual is acting as a consumer or as a producer. Consumers have become so choosy that they no longer will simply accept whatever is offered in the marketplace. In cases where social dictum directs consumers to behave in ways not conducive to sustainability, gradual changes must be facilitated. If necessary, an acquired taste must be developed to like and accept the products of local industry. To facilitate consumer acceptance, the quality of industrial products must be improved to competitive standards. In the past, consumers were expected to make do with the inherent characteristics of products, regardless of potential quality and functional limitations. This has changed drastically in recent years. For a product to satisfy the sophisticated taste of the modern consumer, it must exhibit a high level of quality and responsiveness to the needs of the consumer with respect to global expectations. Only high-quality products and services can survive the prevailing market competition and, thus, fuel the enthusiasm for further sustainability efforts. Some of the approaches for preparing a society for sustainability changes are listed as follows:

- Make changes in small increments
- Highlight the benefits of sustainability development
- Keep citizens informed of the impending changes
- Get citizen groups involved in the decision process
- Promote sustainability change as a transition to a better society
- Allay the fears about potential loss of jobs due to new sustainability programs
- Emphasize the job opportunities to be created from sustainability investments

Addressing the aforementioned issues means using a systems view to tackle the various challenges of executing sustainability. As has been discussed in the preceding sections, the concept of sustainability is a complex one. However, with a systems approach, it is possible to delineate some of its most basic and general characteristics. For our sustainability purposes, a system is simply defined as a set of interrelated elements (or subsystems). The elements can be molecules, organisms, machines, machine components, social groups, or even intangible abstract concepts. The relations, interlinks, or "couplings" between the elements may also have very different manifestations (e.g., economic transactions, flows of energy, exchange of materials, causal linkages, and control pathways). All physical systems are *open* in the sense that they have exchanges of energy, matter, and information with their environment that are significant for their functioning. Therefore, what the system "does," in its behavior, depends not only on the system itself but also on the factors, elements, or variables coming from the environment of the system. The environment impacts "inputs" onto the system, while the system impacts "outputs" onto the environment. This, in essence, is the systems view of sustainability.

CONCLUSIONS

Sustainability implies repeatability, the ability to replicate success again and again to create an enduring pattern of achieving the end goal. Repeatability is not limited to the attributes of the end product. It must also be pursued and practiced in all facets of operations from a systems perspective. It is only through a systems view that we can address the multitude of issues and factors associated with sustainability.

REFERENCES

Agustiady, Tina and Badiru, Adedeji B. (2013), *Sustainability: Utilizing Lean Six Sigma Techniques*, Taylor & Francis CRC Press, Boca Raton, FL.

Badiru, Adedeji B. (Nov. 2010), "The Many Languages of Sustainability," *Industrial Engineer*, Vol. 42, No. 11, pp. 31–34.

Badiru, Adedeji B. (Jan. 2019), "Our Greatest Grand Challenge: To Address Society's Urgent Problems, Engineers Need to Step Up to the Political Plate," *ASEE PRISM*, Vol. 28, No. 6, p. 56.

Industrial Engineering for Sustainability

2

> "Sustainable success lies in what we do rather than what we say."
>
> — *Adedeji Badiru*

INTRODUCTION: EFFICACY OF INDUSTRIAL ENGINEERING IN SUSTAINABILITY

Mask on! Sustainability is at risk, due to the emergence of the COVID-19 pandemic. We must do and act along the frameworks that enable us to tackle issues of sustainability both locally and globally. It is well known that industrial engineers make things happen.

Industrial engineers design, develop, implement, and improve integrated systems that include people, materials, information, equipment, and energy. Industrial engineers accomplish the integration of systems using appropriate analytical, computational, experimental practices, and social science principles. Industrial engineers find creative and better ways of getting things done, including sustainability. Industrial engineers pursue goals and objectives more efficiently.

The tools and techniques of industrial engineering provide efficacious framework to deliver and maintain sustainability globally. The ready definition of sustainability relates to the environment and climate. However, the ideals of sustainability can exist in a variety of platforms, including the environment, business, climate, political ideologies, and national interfaces. Some notable definitions include the following.

Sustainability is the human preservation of the environment, whether economically or socially, through personal responsibility, management of resources, and maintenance of environmental assets.

Sustainability is the act of humans to preserve the pristine nature of our environment and the nature that surrounds us.

Sustainability involves environmental resource management, which is the management of the interaction and impact of human society on the environment.

Sustainability is the prevention of the depletion of our natural resources in order to maintain an ecological balance.

Sustainability is the ability to retain and preserve the stability of the environment for the sake of future generations.

All of the above definitions, views, and approaches are amenable to the application of industrial engineering. Apart from the environmentally focused definitions of sustainability, there are also business-oriented definitions of sustainability. Being sustainable in manufacturing and any financial institution only increases sales and profitability by reducing costs. Doing it is easy and involves awareness and education, policies and programs, and strategic planning.

Sustainability can mean different things to different groups at different times in different places. For example, the pungent smell of manure may signal a desirable blessing in a farmer's environment for agricultural needs. By contrast, the same smell may be abhorrent to city dwellers. Sustainability exists within the system of perception for each person or group.

Critical thinking by using safety and practicality techniques can transform our communities and global environments while moving into the next centuries. Sustainability is based on survival and well-being which is revolved around common sense principles. Understanding our environment and establishing better practices utilizing continuous improvement principles will help us economically and help the generations to come. The topic sounds overwhelming, but utilizing Lean Six Sigma principles will make it a universal way of improving our lives.

When determining what to do to be sustainable, the following four main concepts should be sought after:

- Economical
- Personal
- Societal
- Environmental

Economical terms are related to profitability in the economy. Personal means of sustainability are the effects that are subjected to an individual. Societal concepts deal with informal social gatherings of groups organized by something in common. Finally, the environment is the setting, surroundings, or conditions in which living objects operate. These four concepts unite together for sustainability because of the parallel relationship they have to one another.

The following basic questions should be asked first:

- What needs to be done?
- What can be done?
- What will be done?
- Who will do it?
- When will it be done?
- Where will it be done?
- How will it be done?

The aspects that are critical to sustainability are the most basic ones. These include the environment; natural resources, energy, air, water, and the proper use of them; and finally the understanding of physics, chemistry, biology, and geology. As humans, we must be economical with the natural resources around us. Many may ask, "Why should I protect my environment, and why should I be economical." The answer is simple, it is for well-being. This well-being can be of an individual, resources, work, proper living manners, and pure satisfaction. Without being sustainable, our environment will suffer from aspects such as pollution, not having enough renewable resources especially essential ones, and livelihoods.

The same aspects that are thought about in everyday life for our own well-beings should be thought about for work purposes as well. Emissions are powerful and must be reduced to be successful. Some companies are mandated to only pollute a certain percentage into the environment. Recall, being sustainable in manufacturing only increases sales and profitability by reducing costs. The quality of products is always increased when sustainability is built in. Sustainability also means predictability or doing the same right thing every time. When costs are cut, quality should not be altered or the increase in sales will decrease drastically. The processes behind the quality are the most critical ones in order to be sustainable. It should be sought after to have value-added tasks with quality products. When customers see that sustainability is not being met, the customers will go elsewhere. This also happens when the customers have expectations and are surprised with the product each different time they purchase the product. If the product meets the expectations of customers, the customers will come back. Every time they receive a product that is different from what they expect, they get disappointed and they turn to a different supplier. Not only do the customers go elsewhere, but they complain to their friends and family about their dissatisfaction. Their friends and family now complain to other friends and family about the story they heard, and the company loses sales and hence profitability by sacrificing their quality.

The main global issue in companies is the wastes they incur. The eight main wastes are the following:

1. Transportation
2. Inventory
3. Motion
4. Waiting
5. Overprocessing
6. Overproduction
7. Defects/rework
8. Underutilization of employees

The first five points are discussed below.

Transportation: Materials or parts transferred within the facility should be minimized so that it is in the most strategic manner possible. If a part is coming from shipping and going to manufacturing, the shipping dock must be as close to the manufacturing dock as possible. When each part is utilized, the parts being put in the product should all be in similar vicinities so that there is no waste in transporting the parts back and forth. Also, when the part is completed with its assembly or manufacturing, the packaging area and storage area should be close by or an ease of transportation must occur to limit this type of waste.

Inventory: Inventory should always be looked at as dollar signs on the shelves of where the inventory sits. The raw materials, in-process items, and finished goods are not value-added items when just sitting in a location. Just-in-time (JIT) methodologies should be used to reduce having excess inventory.

JIT is a value-streaming methodology created by Taiichi Ohno at Toyota in the 1950s. The production and delivery of proper items at proper times is the philosophy for JIT to be successful. Upstream activities must occur before downstream activities, aiming for single-piece flow.

Flow, pull, standard work, and takt are the key elements. Changeovers need to be minimized in order for upstream manufacturing processes to have minimal parts until the downstream manufacturing process is ready. This prevents accumulation of excess stock. Scheduling must be effective and consistent, for this process to be maintained. Over exceeding production will also cause inventory issues. Demand of customers' needs is to be measured before producing.

Flow is looked at in order to enhance efficiency. Value-added tasks should be looked for even when difficult to analyze. Many employees find tasks such as cleaning to be value added when in fact they are not. The only tasks that add value are the tasks that the customer would be willing to pay for. The focus should be on the end product and what the customer wants. If the customer is buying a chocolate cake with chocolate frosting, they only want to pay for this cake. They do not want to pay for the changeover between strawberry and chocolate that has to occur before they get their chocolate cake.

Standardized work, also known as standard operating procedure, consists of a definitive set of work procedures with all tasks organized optimally to meet customer needs. Standardization of the process and the activities allow for consistent times and completion of entire processes by all employees at all times and at all circumstances.

Correct locations of materials, parts, equipment, and so forth also help the flow of materials and processes. 5S is a key tool used when determining flow. The 5S will be discussed further in the book.

Motion: Any wasted time while performing any activity is considered a waste of motion. The topic is similar to what was discussed earlier. The customer only wants to pay for value-added tasks. The customer does not want to pay excess labor fees, which are inherent when an employee is unorganized and looking for materials. Any type of wasted motion such as double-handling ingredients, walking further than needed, and stacking multiple times are considered wastes of motion.

Waiting: When any person is waiting for tools, materials, equipment, or other personnel, they are being non-value added. JIT should again be used, yet this time with people as the driving factor. The next step in the process should come just in time for the next person to be able to act on that process without wait time occurring. If wait time is built in the process, there is excess time built into the process.

Overprocessing: A question many people ask when doing tasks is, "Why do you do it that way?" The common answer is, "That's how we have always done it," or "That's what the procedure says." What should be thought about is if the process has too much work built into it. If a process requires 5 minutes of stirring, it should be asked if the materials would be the exact same after 3 minutes of stirring. Minimizing excess processing saves time and eliminates waste while increasing efficiency and productivity. This type of thinking requires "Thinking Outside of the Box."

The following is the strategy for utility cost reduction:

- Utility Cost = Utility Usage × Utility Price
- Electricity $/Year = kWh/Year × $/kWh
- Natural Gas $/Year = Cu. Ft./Year × $/Cu. Ft.
- Water $/Year = Gallons/Year × $/Gallon

Simply put, utility cost reductions are the product of utility usage reductions and utility price reductions. Focusing entirely on usage reductions or entirely on price reductions is not a well-balanced strategy.

In order to correctly estimate utility usage and cost savings, two estimates must be prepared. Annual utility usage savings and cost savings are the difference between the "before" usage/cost estimate and the "after" usage/cost estimate:

• Annual "Before-Retrofit" Utility Usage /Cost − Annual "After-Retrofit" Utility Usage/Cost = Annual Utility Usage/Cost Savings

Because this is a multifactor analysis, e.g., before and after levels of equipment loading/efficiency/demand, before and after annual hours of operation, before and after utility prices, and so on, there are lots of opportunities for errors. Thus, the uncertainty of the utility usage and cost reduction estimates must be taken into account. The conservative approach is to discount the estimates by the associated level of uncertainty.

Utility cost reduction measures and focuses for sustainability are as follows:

1. Efficient operation and effective maintenance of utility-consuming equipment
2. Competitive procurement of utilities
3. Cost-effective expense improvements
4. Cost-effective capital projects and retrofits
5. Efficient design of new buildings and plant expansions

The following are examples through simple means of sustainability:

• Computers, radios, stereo systems should be turned off when not in use and unplugged when possible
• Utilize programmable thermostats to save up to 20% annually
• Turn off electricity in rooms not being utilized − think about freezers in factories that are not being used
• Focus on:
  • Electricity − #1 source of resource consumption
  • Natural gas − #2 source of resource consumption
  • Water/sewer − #3 source of resource consumption

Utility costs are one of the largest categories in the annual expense budget for facilities and maintenance organizations. Unlike property taxes and depreciation costs, utility costs can be readily reduced.

Preventative maintenance is 67% to 75% less expensive than repair-upon-failure, so it is a measure that reduces net expenses concurrent with implementation. Preventative maintenance need not be applied wholesale – it can be implemented in stages and/or for selected categories of assets.

The following are the eight best practices for improved energy efficiency:

1. Increase the efficiency of all motors and motor-driven systems
2. Improve building lighting
3. Upgrade heating, ventilating, and cooling systems
4. Capture the benefits of utility competition
5. Empower your employees to do more
6. Use water-reduction equipment and practices
7. Explore energy savings through increased use of the Internet
8. Implement comprehensive facility energy and environmental management

The next topic discusses the areas of sustainability with particular aspects of operational efficiency.

AREAS OF SUSTAINABILITY

When discussing the different areas of sustainability, particular aspects come to mind (Agustiady and Badiru 2013). Sustaining the environment is first and foremost, but we need to understand what is included in the environmental aspects. The most common ways to sustain the environment is through energy, water, and agricultural conservation. To do this, we must be well educated in sustainability for it to be implemented and have lasting results. First, we need to understand energy consumption. Energy is defined as the capacity of a physical system to perform work through heat, kinetics, mechanical systems, light, or electrical means.

The beginning process of saving energy comes from simple means such as turning lights and any sources that provide power. Computers, radios, stereo systems, and the like should be not only turned off when not in use but also unplugged when possible. Surge protectors are good ways of being able to ease into this process. Multiple items can be plugged into the surge protector and the switch can simply be turned off when not in use.

Many thermostats are programmable now so that the air conditioning or heat can be turned to a lower/higher temperature when you are off at work or elsewhere. The trick to programmable thermostats is that it gradually gets back to the temperature of normal operating conditions when you enter back in the home. These savings can be up to 20% annually if utilized properly. Another great way to save energy in your home is to close any vents to rooms that are underutilized. The extra heat or air will come into the other rooms saving costs again.

Weather proofing windows with weather strips or plastic insulation will also help drafts or leaks. Curtains can also help with this process and currently insulated curtains are available to keep in heat or air. Simply insulating homes can protect excess heat or air costs and has a quick payback of normally less than one year.

Microwaves or small toaster ovens use much less energy than stoves or ovens. These should be used when cooking or defrosting small amounts of food.

Keeping the cold energy in the refrigerator is important because of the amount of energy it takes. Therefore, closing the fridge as soon as possible is very important. Some fridges come with energy-saving signals, such as beeps, to warn you that the fridge has been open for longer than the allotted time. Remember that old fridges also account for a large amount of energy.

Batteries also take up energy that can be saved by turning off any toys, games, and the like. Rechargeable batteries can save energy and reduce the toxics of the dangerous heavy metals that are emitted from batteries when thrown away. Batteries should only be disposed in toxic waste disposals.

Water is the next critical area where sustainability is considered. The basics of turning off water whenever possible are the key to the savings. Most cities use the most amount of energy supplying water and cleaning up the water after it has been used. Turning off water when brushing teeth or when soaping dishes is simple an idea. Taking shorter showers is another way to save on the use of hot water. Hot water heaters account for almost one-fourth of the home's energy usage.

Surprisingly, dishwashers save up to 40% more water than hand-washing dishes as well! If hand washing is a necessity, the best way to save is to fill up one side of the sink with clean soapy water and rinsing on the other side.

Using your own energy versus power sources will help in energy saving as well. Examples such as using rakes versus leaf blowers or push mowers versus gas mowers saves not only energy but helps your own human well-being!

Anything that can be thrown away should be thought about before purchasing. Think about disposable products such as plastic cups, plastic bags, napkins, and paper plates. Whenever a reusable source can be used, it should be used. Cloth napkins can cut down on the excessive usage of paper napkins. Also, the plastic bags given at grocery stores normally end up in the trash. Consider taking your own bags with you to recycle the use. Newspapers should always be recycled; 36 million trees a year could be saved if everyone recycled their newspaper every day.

If these simple things can be done quickly and efficiently in one's home, think about the benefits in a corporation for these savings. The benefits can have major savings and the paybacks for buying energy-efficient bulbs, surge protectors, automated light sensors, and the like will be immediate.

The beginning programs for these sustainability acts are simple. They consist of the following:

- Awareness
- Energy-efficient coding
- Communications programs
- Green schools and offices programs
- Energy alliances programs
- Industrial programs
- Policy and research programs
- Water saving acts
- Commercial building consortiums

Awareness begins with education. Public awareness, training, and classroom activities will help the society move toward sustainable actions for the environment. Attainability becomes an argument when discussing sustainability. The lack of agreement behind these acts should be followed up with facts behind sustainability and will aware others of truths behind the matter. Instead of focusing on what is wrong with our society and how difficult it is to maintain these sustainable measures, we should begin with the quick fixes and how simple ideas can help the environment. If each person starts doing just a small part, the effects can be immense. The small measures will then lead to much larger projects, but the start has to begin somewhere. The big differences between the reality of being sustainable as a culture and the education behind being sustainable should be known. Education will lead the environment from being protected in small pieces while slowly investing in the ideas.

Education will affect the following three main areas of planning:

- Implementation
- Decision making
- Quality of life

Without educating the work force to keep them informed, the implementation of sustainability cannot be founded. Highly illiterate forces have a difficult time with developmental plans and actions. Instead, these types of areas and people will have to buy forms of energy and goods spending a great deal of money.

Decision making is important because without factual and motivational information, decisions cannot be made. Development options with educational topics will help people become skilled and technical and will also lead the act of persuasion. The persuasion is simple due to having educational information that leads to a better society because the facts protect the environment and social structure.

Finally, the quality of life is important to each individual person. The education of these practices leads to higher economic statuses, improvement of life conditions, and the future of coming generations. The quality of life helps both individuals and national geography.

Sustainable development plays a key role during this educational phase. The concept behind sustainable development is still evolving. The definition of sustainable development is the concept of needs mostly from the idea of limitations imposed by society and technology to prevent the environment from meeting present and future needs. The Brundtland Commission is to be credited for the original description of sustainable development. They quoted, "Sustainable development is development that meets the needs of the present without compromising the ability of future generations to meet their own needs."

This concept is thought to have the following three main components:

- Environment
- Society
- Economy

These acts need to be combined and not seen separately. The thought is to have a balance of all of these in order to have a healthy environment with proper resources for its citizens in the cleanliest manner possible.

Remember that the environment is the setting, surroundings, or conditions in which living objects operate. Society is the collective group of people living together in a particular region with the same customs, laws, or organizations. The economy is the prosperity, possessions, and resources of a country or region in terms of production and consumption of goods and services. They can all be tied together because they need each other to depend on.

Many principles are tied in with sustainable development. The definitions have been established from "The Rio Declaration on Environment and Development" and the 18 points are listed below:

- People are entitled to a healthy and productive life in harmony with nature.

- Development today must not undermine the development and environment needs of present and future generations.
- Nations have the sovereign right to exploit their own resources, but without causing environmental damage beyond their borders.
- Nations shall develop international laws to provide compensation for damage that activities under their control cause to areas beyond their borders.
- Nations shall use the precautionary approach to protect the environment. Where there are threats of serious or irreversible damage, scientific uncertainty shall not be used to postpone cost-effective measures to prevent environmental degradation.
- In order to achieve sustainable development, environmental protection shall constitute an integral part of the development process and cannot be considered in isolation from it. Eradicating poverty and reducing disparities in living standards in different parts of the world are essential to achieve sustainable development and meet the needs of the majority of people.
- Nations shall cooperate to conserve, protect, and restore the health and integrity of the Earth's ecosystem. The developed countries acknowledge the responsibility that they bear in the international pursuit of sustainable development in view of the pressures their societies place on the global environment and of the technologies and financial resources they command.
- Nations should reduce and eliminate unsustainable patterns of production and consumption and promote appropriate demographic policies.
- Environmental issues are best handled with the participation of all concerned citizens. Nations shall facilitate and encourage public awareness and participation by making environmental information widely available.
- Nations shall enact effective environmental laws and develop national law regarding liability for the victims of pollution and other environmental damages. Where they have authority, nations shall assess the environmental impact of proposed activities that are likely to have a significant adverse impact.
- Nations should cooperate to promote an open international economic system that will lead to economic growth and sustainable development in all countries. Environmental policies should not be used as an unjustifiable means of restricting international trade.
- The polluter should, in principle, bear the cost of pollution.
- Nations shall warn one another of natural disasters or activities that may have harmful transboundary impacts.

- Sustainable development requires better scientific understanding of the problems. Nations should share knowledge and innovative technologies to achieve the goal of sustainability.
- The full participation of women is essential to achieve sustainable development. The creativity, ideals and courage of youth, and the knowledge of indigenous people are needed too. Nations should recognize and support the identity, culture, and interests of indigenous people.
- Warfare is inherently destructive of sustainable development, and nations shall respect international laws protecting the environment in times of armed conflict and shall cooperate in their further establishment.
- Peace, development, and environmental protection are interdependent and indivisible.

These "Rio Principles" are clear parameters for the vision of the future of the world in order to be developmentally sustainable with specific abstracts of factual concepts.

Energy-efficient coding are the codes that set minimum requirements for energy-efficient design and construction for new and renovated buildings that impact energy usage. The objective is to have consistent and long-lasting results for energy saving. The benefits will help the environment both tangibly and intangibly. The reduction in energy helps the pollution concept by reducing environmental pollutions. There are direct savings to and financial benefits due to a decrease in energy consumption and an increase in energy-efficient technologies, which will also lead to economic opportunities for the businesses. Once a building is already constructed, it is difficult to achieve the energy efficiencies without retrofitting the building. Since buildings stay up for decades, the energy coding should be done at the beginning stages for maximum benefits.

Contemporary energy codes can save up to 330 trillion BTU by 2030, which is almost 2% of the total current residential energy consumption. The states can take the lead in the energy-saving initiatives by mandating energy coding in new or retrofitted buildings based on data from US Department of Energy website: www.doe.gov.

Communication programs for sustainability include basic marketing techniques. Commercials, t-shirts promoting energy-saving activities, and mobile media are the most common ways to "spread the word." Programs include seminars to raise awareness and scholarship grants to reward students with bright ideas. Finally, donations help the sustainable efforts for the communication plans for the future.

Green Schools and Offices programs comprise many communities that take initiatives to have only green products. The communication of where and what to get to be green is communicated via directories and online sources. Green Schools and Offices programs give hazardous material warnings of items such as PVC plastic and give the benefits of recycling programs. These programs are meant to show the community their commitment in the matter by safeguarding health, saving money, raising test scores through better air quality and environments, and empowering kids.

Energy alliance programs give tips and resources for making energy efficiency easy and affordable. There are many community-based nonprofit organizations that provide property owners information on what to do, who to call, how to pay for resources, and what the proper tactics are for energy saving. Programs such as these are not associated with any political parties and are solely performed for better environments. The programs have also begun to involve renters, contractors, and business partners.

Industrial programs perform similar techniques for energy saving by having steering committees to gain executive buy-in from the beginning and create sustainable goals and establishments with measurable results. Industrial programs involve best practices by exploring company motivations for the program that have had successive results. The marketing research is a key aspect to the successes for industrial programs. The goal of industrial programs is to have clean air and transportation, upcoming technology, corporate sustainability, and a means for building efficiency sectors.

Policy and research programs involve sustainable-science-oriented programs that improve science through better awareness of human-environment systems. The knowledge behind the policies increases the ease of implementation by having data-driven results for the changes to be implemented. The promotion for sustainability includes concepts to increase knowledge, train students and faculty, and continuous education through teaching and outreach.

There are several water-saving acts that involve goals for reduction in urban water usage by 20% in eight to ten years. Incremental changes are to be taken with a goal of having at least 10% of savings in the next five years. Water management plans encompass giving deliverables on the amount of water usage acceptable through technical methodologies with strict criteria. The water-saving act has a goal to not approve any retail water suppliers to gain any grants or loans if their water conservation requirements are not met. There are plans for an adoptive pricing structure for water by basing it on parts or quantities delivered. Efficient management strategies are to be encompassed in this water-saving act.

Commercial building consortiums are developed from the US Department of Energy to gain building asset rating systems that can be built for new and existing commercial buildings to provide information to the stakeholders. The concept is due to the loss of heating and cooling by up to 30% from commercial buildings, which is through the poor use or insulation of doors, windows, curtains, skylights, which is also known as fenestration. The goal behind commercial building consortiums is to have all building sectors in the United States utilize zero energy through the implementation of aggressive energy-saving mechanisms to reduce demand of energy by 70% to 80% while still maintaining energy requirements through renewable resources.

These different areas of sustainability help benchmark the future use of energy by primarily using the domination of communication, knowledge, and education as a means of well-being for future generations.

WHAT HAPPENS GLOBALLY IF WE ARE NOT SUSTAINABLE?

According to the US Census, there are approximately 312 million people in the United States and approximately 6.9 billion people in the world. With this many people, there is a fear of running out of natural resources especially energy. Renewable sources of energy are needed to be sustainable. Ultimately, all living organisms require energy, and if they utilize energy without reproducing it in some way, the energy WILL run out. Revitalizing all forms of energy must be accomplished to be sustainable. This means, reorganize, restore, and differentiate. If we continue to use resources without reinvesting, all energy will be lost and there will be no hope for future generations. Social beings working together can help current and future generations by sustaining positive relationships through education and understanding of laws and practices. Locally purchasing ingredients and goods also helps the environment so that there are no monopolies. Monopolies are defined as one company or industry controlling the services or goods provided. With the concept of monopoly, no other beings are given a chance to be held socially responsible. The consequences include not having hope for future generations to be successful because no bartering can be done from one business or organization to another.

Overconsumption can come from pressures of being known as successful or prestigious. However, having the best cars, homes, and things will lead to holding others at a pace they cannot keep up with. According to *The Story of Stuff*, 2007, we have consumed 30% of Earth's natural resources in the past 30 years. Only a few of these resources can be replenished. Only 20% of old-growth forest is remaining and 75% of fisheries are producing at or above capacity. The United States is accountable for the most abusive global consumer with only 5% of the world population and 30% of the worldwide consumption. About 99% of raw materials are discarded without being recycled. This will be catastrophic if these standards are maintained or worsened over the years.

The food chain circumstance that animals use can be converted to humans with this mannerism, meaning the best or most successful will only come out on top. Instead, all philosophies should be investigated or educated upon before making decisions. Values can then be set with the proper principles in mind-holding relationships as important aspects. This is another form of unselfishness that should be practiced in order to stay globally friendly and sustainable. These concepts will not only help current practices and environments but even better hopes for future generation after generation. If we keep consuming resources and just moving to different areas to find more resources, eventually the resources in all areas will run out leaving the world completely out of resources.

CONCLUSIONS

This chapter has presented the various views, definitions, and perceptions of sustainability within the context of business and industry. The chapter emphasizes the possible deleterious effects of not pursuing sustainability. Although the most common response to the call for sustainability relates to environmental sustainability, the fact is that the ideals of sustainability exist in more platforms than the environment.

REFERENCE

Agustiady, Tina and Badiru, Adedeji B. (2013), *Sustainability: Utilizing Lean Six Sigma Techniques*, Taylor & Francis CRC Press, Boca Raton, FL.

The 14 Grand Challenges for Engineering

3

> "Life is lived and sustained through new challenges."
>
> — *Adedeji Badiru*

INTRODUCTION: NATIONAL ACADEMY OF ENGINEERING: THE 14 GRAND CHALLENGES FOR ENGINEERING

The premise of this chapter, in consonance with the theme of this book, is to identify and highlight how the tools and techniques of industrial and systems engineering can be brought to bear on global challenges (Agustiady and Badiru 2019; Badiru 2010, 1). The National Academy of Engineering (NAE), in 2008, released a list of the 14 grand challenges for engineering in the coming years. Each area of challenges constitutes a complex project that must be planned and executed strategically. The 14 challenges, which can be viewed as science, technology, engineering, and mathematics (STEM) areas, are listed as follows:

1. Make solar energy affordable
2. Provide energy from fusion
3. Develop carbon sequestration methods
4. Manage the nitrogen cycle
5. Provide access to clean water
6. Restore and improve urban infrastructure
7. Advance health informatics

8. Engineer better medicines
9. Reverse-engineer the brain
10. Prevent nuclear terror
11. Secure cyberspace
12. Enhance virtual reality
13. Advance personalized learning
14. Engineer the tools for scientific discovery

The aforementioned list has sustainability written all over it, in one form or another. In fact, NAE arrived at the 14 topics through an international group of leading technological researchers and practitioners, who surveyed to identify the grand challenges for engineering in the 21st century. The result was the 2008 14 game-changing goals for improving life on the planet. The survey committee suggested that the 14 grand challenges be categorized into four cross-cutting themes, encompassing sustainability, health, security, and joy of living.

The list of existing and forthcoming engineering challenges indicates an urgent need to apply comprehensive systems-based project management to bring about new products, services, and results efficiently within cost and schedule constraints, possibly under the umbrella of sustainability. Project management, executed within the scope of industrial engineering, can effectively be applied to the grand challenges to ensure a realization of the objectives.

Although the NAE list focuses on engineering challenges, the fact is that every item on the list has the involvement of general areas of STEM, in one form or another. The STEM elements of each area of engineering challenge are contained in the following definitions. Industrial engineering cuts across other engineering disciplines. Thus, there is a potential for collaborative applications in every topic included in the following list.

1. **Advance Personalized Learning**

   A growing appreciation of individual preferences and aptitudes has led toward more "personalized learning," in which instruction is tailored to a student's individual needs. Given the diversity of individual preferences, and the complexity of each human brain, developing teaching methods that optimize learning will require engineering solutions of the future. The emergence of COVID-19 has made this topic even more relevant and urgent due to the need to embrace remote learning.

2. **Make Solar Energy Economical**

   As of 2008, when the list was first published, solar energy provided less than 1% of the world's total energy, but it has the potential to provide much more. There has been some increase in the percentage of solar usage, but there is room for more. We need to continue our systems-driven efforts in this endeavor.

3. **Enhance Virtual Reality**

Within many specialized fields, from psychiatry to education, virtual reality is becoming a powerful new tool for training practitioners and treating patients, in addition to its growing use in various forms of entertainment. Mixed-mode simulation of production systems has been practiced by industrial engineers for decades. The same operational wherewithal can be applied to enhancing virtual reality in the context on the new world order in business, industry, academia, and government.

4. **Reverse-Engineer the Brain**

A lot of research has been focused on creating thinking machines – computers capable of emulating human intelligence – however, reverse-engineering the brain could have multiple impacts that go far beyond artificial intelligence and will promise great advances in health care, manufacturing, and communication. The same technical principles of reverse-engineering, which business and industry already use, can be applied to human anatomy and physiological challenges.

5. **Engineer Better Medicines**

Engineering can enable the development of new systems to use genetic information, sense small changes in the body, assess new drugs, and deliver vaccines to provide health care directly tailored to each person. The intense worldwide search for a vaccine for COVID-19 comes to mind in this regard. We need ways to engineer and develop better medicines quickly, effectively, and efficiently. Such a process is the bastion of industrial engineers, who can work with a diverse collection of scientists to expedite and secure production processes. Industrial engineering tools, such as Lean production and Six Sigma, are very much needed for engineering better medicines.

6. **Advance Health Informatics**

As computers have become available for all aspects of human endeavors, there is now a consensus that a systematic approach to health informatics – the acquisition, management, and use of information in health – can greatly enhance the quality and efficiency of medical care and the response to widespread public health emergencies. The rapid emergence of new disciplines in data analytics and data science is in line with this stated area of need included in the 14 grand challenges.

7. **Restore and Improve Urban Infrastructure**

Infrastructure is the combination of fundamental systems that support a community, region, or country. Society faces the formidable challenge of modernizing the fundamental structures that will support our civilization in centuries ahead. Industrial engineers have facilities design and urban infrastructure planning in their skill sets. Thus, there is an alignment between this area of urgent need and the ready capabilities of industrial engineers.

8. **Secure Cyberspace**

   Computer systems are involved in the management of almost all areas of our lives – from electronic communications and data systems to controlling traffic lights to routing airplanes. It is clear that engineering needs to develop innovations for addressing a long list of cybersecurity priorities. The increasing threat to cyberspace and the risk of unexpected hacking and authorized data incursion have made it imperative to expedite research and development in cyberspace security.

9. **Provide Access to Clean Water**

   The world's water supplies are facing new threats; affordable, advanced technologies could make a difference for millions of people around the world. Efficient management of resources is an area of interest and expertise for industrial engineers. Water is a resource that we often take for granted. With a systems-oriented view of all resources, industrial engineers can be instrumental in the global efforts to provide access to clean water.

10. **Provide Energy from Fusion**

    Human-engineered fusion has been demonstrated on a small scale. The challenge is to scale-up the process to commercial proportions in an efficient, economical, and environmentally benign way. This is a highly technical area of pursuit. But it is still an area that can benefit from better managerial policies and procedures from the perspectives of industrial and systems engineering. No matter how technically astute a system might be, it will still need good policies and collaborative governance. This is something that industrial engineers can contribute to in the pursuit of providing energy from alternate sources.

11. **Prevent Nuclear Terror**

    The need for technologies to prevent and respond to a nuclear attack is growing. The threat for a nuclear attack is always human-driven. The more we understand how the other side thinks and operates, the better we can handle on how to prevent nuclear terror. The human-focused practice of industrial engineering may have something to offer with respect to understanding human behavior, reactions, and tendencies. Both preparation and prevention are essential in the endeavor to better leverage nuclear capabilities while limiting the chances for terror.

12. **Manage the Nitrogen Cycle**

    Engineers can help restore balance to the nitrogen cycle with better fertilization technologies and by capturing and recycling waste. Managing the nitrogen cycle requires management, from the standpoint of technical requirements and human sensitivities. The reuse, recover, and recycle programs that industrial engineers often participate in can find a place in this topic.

13. **Develop Carbon Sequestration Methods**

    Engineers are working on ways to capture and store excess carbon dioxide to prevent global warming. Industrial engineers are already working within the domain of designing efficient storage systems, whether for industrial physical assets or non-physical resources. It is all about the strategic placement of things, while considering the systems influence coming from other things.

14. **Engineer the Tools of Scientific Discovery**

In the century ahead, engineers will continue to be partners with scientists in the great quest for understanding many unanswered questions of nature. The toolbox of industrial engineers contains what it takes to advance and leverage the tools and processes for scientific discovery.

Our society will be tackling these grand challenges for the foreseeable decades, and project management is one avenue through which we can ensure that the desired products, services, and results can be achieved. With the positive outcomes of these projects achieved, we can improve the quality of life for everyone and our entire world can benefit positively. In the context of tackling the grand challenges as system-based projects, some of the critical issues to address are as follows:

Strategic implementation plans
Strategic communication
Knowledge management
Evolution of virtual operating environment
Structural analysis of projects
Analysis of integrative functional areas
Project concept mapping
Prudent application of technology
Scientific control
Engineering research and development

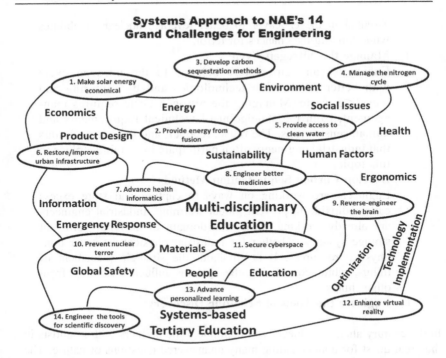

FIGURE 3.1 Semantic Network of Systems-Based Approach to the 14 Grand Challenges

THE GRAND CHALLENGES WITH OVERLAPPING INTEGRATION

We must integrate all the elements of a project on the basis of alignment of functional goals. Systems overlap for integration purposes can conceptually be represented as projection integrals by considering areas bounded by the common elements of subsystems. Multidisciplinary education is essential in grasping the integrated concepts and principles that exist among the elements of the 14 grand challenges for engineering. Figure 3.1 presents a semantic network of how the challenges intertwine and cross paths with industrial engineering tools and techniques.

CONCLUSIONS

In terms of a summary for this chapter, systems integration is the synergistic linking together of the various components, elements, and subsystems of a system, where the system may be a complex project, a large endeavor, or an expansive organization. Activities that are resident within the system must be managed from both the technical and the managerial standpoints. Any weak link in the system, no matter how small, can be the reason for the overall system failure. In this regard, every component of a project is a critical element that must be nurtured and controlled. Embracing the systems principles for project management will increase the likelihood of success of projects.

REFERENCES

Badiru, Adedeji B. (Jan 2019), "Our Greatest Grand Challenge: To Address Society's Urgent Problems, Engineers Need to Step Up to the Political Plate," *ASEE PRISM*, p. 56.

Badiru, Adedeji B. (Nov 2010), "The Many Languages of Sustainability," *Industrial Engineer*, Vol. 42, No. 11, pp. 31–34.

Agustiady, Tina and Badiru, Adedeji B. (2013), *Sustainability: Utilizing Lean Six Sigma Techniques*, Taylor & Francis CRC Press, Boca Raton, FL.

DEJI Systems Model for Sustainability Innovation

4

"The bridge to sustainability is the systems framework."

— Adedeji Badiru

INTRODUCTION

"A new way of doing things" is one of the definitions of innovation. Sustainability requires new ways of doing things. Whether it is this definition or any other definition, sustainability innovation must have a buy-in from stakeholders and it must be integrated into the operating environment. This chapter presents the application of the trademarked Design, Evaluate, Justify, and Integrate (DEJI) systems model[®] to the management of sustainability innovation processes. The model provides a generic pathway for sustainability innovation design, evaluation, justification, and integration. The DEJI systems model (Badiru 2012, 2019) is a good tool for ensuring that the proposed innovation fits the operating environment of the organization, the community, or the nation. The DEJI systems model is applicable for innovation design, evaluation, justification, and integration. Figure 4.1 illustrates the overall framework of DEJI systems model (more details on the model can be obtained from www.DEJImodel.com).

FIGURE 4.1 Elements of the DEJI Systems Model for Innovation.

WHY THE DEJI SYSTEMS MODEL?

Driven by analytical tools, technical professionals tend to want to jump to the coordination and implementation stage of a project. That is, jumping to the design functionality stage while dispensing with intermediate steps, where nontechnical and "soft" issues might exist. However, those intermediate stages are often more critical for systems success rather than the pure analytical foundation. Items such as needs analysis, gap analysis, user involvement, communication, cooperation, resource requirement analysis, budget flow, leadership support, workforce acceptance, project desirability, and so on, are essential before getting to the end point of the project. This is the essential narrative that highlights the efficacy of the approach of DEJI systems model, which takes a project sequentially through the stages of design, evaluation, justification, and, finally, integration. These step-by-step stages allow important considerations, technical or otherwise, to be addressed in the project. It is of utmost importance to understand how the proposed product of the project will integrate with and align with the existing organizational framework. DEJI systems model makes it imperative to do an *a-priori* evaluation of the potential impact that the project output might have on the prevailing environment.

In implementation, the model can be customized for specific needs related to sustainability. Explanations and examples for design, evaluation, justification, and integration are provided throughout the chapter.

Several factors relating to sustainability innovation are amenable to the application of the DEJI systems model. Some of these are discussed in the sections that follow. Wherever innovation is mentioned in this chapter, the specific focus is on sustainability innovation.

SUSTAINABILITY INNOVATION QUALITY MANAGEMENT

Sustainable "sustainability" requires innovation, which can come in various modes and flavors, technical or nontechnical, each requiring a measure of quality management. Quality is a measure of customer satisfaction and a product's "fit-for-use" status. To perform its intended functions, a product must provide a balanced level of satisfaction to both the producer and the customer. For the purpose of sustainability pursuits, we present the following comprehensive definition of sustainability quality:

> *Quality refers to an equilibrium level of functionality possessed by a product or service based on the producer's capability and the customer's needs.*

Based on the aforementioned definition, quality refers to the combination of characteristics of a product, process, or service that determines the product's ability to satisfy specific needs. Quality is a product's ability to conform to specifications, where specifications represent the customer's needs or government regulations. The attainment of quality in a product is the responsibility of every employee in an organization, and the production and preservation of quality should be a commitment that extends all the way from the producer to the customer. Products that are designed to have high quality cannot maintain the inherent quality at the user's end of the spectrum if they are not used properly.

The functional usage of a product should match the functional specifications for the product within the prevailing usage environment. The ultimate judge for the quality of a product, however, is the perception of the user, and differing circumstances may alter that perception. A product that is perceived as being of high quality for one purpose at a given time may not be seen as having acceptable quality for another purpose in another time frame. Industrial quality standards provide a common basis for global commerce. Customer satisfaction or production efficiency cannot be achieved without product standards. Regulatory, consensus, and contractual requirements should be taken into account when developing product standards driven by innovation. These are described in the following sections.

REGULATORY STANDARDS

This refers to standards that are imposed by a governing body, such as a government agency. All firms within the jurisdiction of the agency are required to comply with the prevailing regulatory standards.

CONSENSUS STANDARDS

This refers to a general and mutual agreement between companies to abide by a set of self-imposed standards.

CONTRACTUAL STANDARDS

Contractual standards are imposed by the customer based on case-by-case or order-by-order needs. Most international standards will fall into the category of consensus standards, simply because a lack of an international agreement often leads to trade barriers.

INDUSTRY AND LEGAL STANDARDS

Along with the aforementioned standards, there are also self-coordinated standards such as industry standards and legal standards.

INNOVATIVE PRODUCT DESIGN

The initial step in any manufacturing effort is the development of a manufacturable and marketable product. An analysis of what is required for a design and what is available for the design should be conducted in the planning phase of a design project. The development process must cover analyses of the product configuration, the raw materials required, production costs, and potential profits. Design engineers must select appropriate materials, the product must be expected to operate efficiently for a reasonable length of time (reliability and durability), and it must be possible to manufacture the product at a competitive cost. The design process will be influenced by the required labor skills, production technology, and raw materials. Product planning is substantially influenced by the level of customer sophistication, enhanced technology, and competition pressures. These are all project-related issues that can be enhanced by project management. The designer must recognize changes in all these factors and incorporate them into the design process. Design project management provides a guideline for the initiation, implementation, and termination of a design effort. It sets guidelines for specific design objectives, structure, tasks, milestones, personnel, cost, equipment, performance, and problem resolutions. The steps involved include planning, organizing, scheduling, and control. The availability of technical expertise within an organization and outside of it should be reviewed. The primary question of whether or not a design is needed at all should be addressed. The "make" or "buy," "lease" or "rent," and "do nothing" alternatives to a proposed design should be among the considerations.

In the initial stage of design planning, the internal and external factors that may influence the design should be determined and given relative weights according to priority. Examples of such influential factors include organizational goals, labor situations, market profile, expected return on design investment, technical manpower availability, time constraints, state of the technology, and design liabilities. The desired components of a design plan include summary of the design plan, design objectives, design approach, implementation requirements, design schedule, required resources, available resources, design performance measures, and contingency plans.

DESIGN FEASIBILITY

The feasibility of a proposed design can be ascertained in terms of technical factors, economic factors, or both. A feasibility study is documented with a report showing all the ramifications of the design. A report of the design's feasibility should cover statements about the need, the design process, the cost feasibility, and the design effectiveness. The need for a design may originate from within the organization, from another organization, from the public, or from the customer. Pertinent questions for design feasibility review include: Is the need significant enough to warrant the proposed design? Will the need still exist by the time the design is finished? What are alternate means of satisfying the need? What technical interfaces are required for the design? What is the economic impact of the need? What is the return, financially, on the design change?

A design breakdown structure (DBS) is a flowchart of design tasks required to accomplish design objectives. Tasks that are contained in the DBS collectively describe the overall design. The tasks may involve hardware products, software products, services, and information. The DBS helps to describe the link between the end objective and its components. It shows design elements in the conceptual framework for the purposes of planning and control. The objective of developing a DBS is to study the elemental components of a design project in detail, thus permitting a "divide and conquer" approach. Overall design planning and control can be significantly improved by using DBS. A large design may be decomposed into smaller subdesigns, which may, in turn, be decomposed into task groups. Definable subgoals of a design problem may be used to determine appropriate points at which to decompose the design.

Individual components in a DBS are referred to as *DBS elements* and the hierarchy of each is designated by a level identifier. Elements at the same level of subdivision are said to be of the same DBS level. Descending levels provide increasingly detailed definition of design tasks. The complexity of a design and the degree of control desired are used to determine the number of levels to have in a DBS. Level I of a DBS contains only the final design purpose. This item should be identifiable directly as an organizational goal. Level II contains the major subsections of the design. These subsections are usually identified by their contiguous location or by their related purpose. Level III contains definable components of the level II subsections. Subsequent levels are constructed in more specific details, depending on the level of control desired. If a complete DBS becomes too crowded, separate DBSs may be drawn for the level II components, for example. A specification of design should accompany the DBS. A statement of design is a narrative of the design to be generated. It should include the objectives of the design, its nature, the resource requirements, and a tentative schedule. Each DBS element is assigned a code (usually numeric) that is used for the element's identification throughout the design lifecycle.

DESIGN STAGES

The guidelines for the various stages in the lifecycle of a design can be summarized in the following way:

1. **Definition of design problem:** Define problem and specify the importance of the problem, emphasize the need for a focused design problem, identify designers willing to contribute expertise to the design process, and disseminate the design plan.
2. **Personnel assignment:** The design group and the respective tasks should be announced and a design manager should be appointed to oversee the design effort.
3. **Design initiation:** Arrange organizational meeting, discuss general approach to the design problem, announce specific design plan, and arrange for the use of required hardware and tools.
4. **Design prototype:** Develop a prototype design, test an initial implementation, and learn more about the design problem from test results.
5. **Full design development:** Expand the prototype design and incorporate user requirements.
6. **Design verification:** Get designers and potential users involved, ensure that the design performs as designed, and modify the design as needed.
7. **Design validation:** Ensure that the design yields the expected outputs. Validation can address design performance level, deviation from expected outputs, and the effectiveness of the solution to the problem.
8. **Design integration:** Implement the full design, ensure the design is compatible with existing designs and manufacturing processes, and arrange for design transfer to other processes.
9. **Design feedback analysis:** What are the key lessons from the design effort? Were enough resources assigned? Was the design completed on time? Why or Why not?
10. **Design maintenance:** Arrange for continuing technical support of the design and update design as new information or technology becomes available.
11. **Design documentation:** Prepare full documentation of the design and document the administrative process used in generating the design.

CULTURAL AND SOCIAL COMPATIBILITY ISSUES

Cultural infeasibility is one of the major impediments to outsourcing sustainability innovation in a wide-open market. The business climate can be very volatile. This volatility, coupled with cultural limitations, creates problematic operational limitations, particularly in an emerging technology. The pervasiveness of online transactions overwhelms the strict cultural norms in many markets. The cultural feasibility of information-based outsourcing needs to be evaluated from the standpoint of where information originates, where it is intended to go, and who comes into contact with the information. For example, the revelation of personal information is frowned upon in many developing countries, where there may be an interest in outsourced innovation engagements. Consequently, this impedes the collection, storage, and distribution of workforce information that may be vital to the success of outsourcing. For outsourcing to be successfully implemented in such settings, assurances must be incorporated into the hardware and software implementations so as to conciliate the workforce. Accidental or deliberate mismanagement of information is a more worrisome aspect of information technology (IT) than it is in the Western world, where enhanced techniques are available to correct information errors. What is socially acceptable in the outsourcing culture may not be acceptable in the receiving culture and vice versa.

ADMINISTRATIVE COMPATIBILITY

Administrative or managerial feasibility involves the ability to create and sustain an infrastructure to support an operational goal. Should such an infrastructure not be in existence or unstable, then we have a case of administrative infeasibility. In developing countries, a lack of trained manpower precludes a stable infrastructure for some types of industrial outsourcing. Even where trained individuals are available, the lack of coordination makes it almost impossible to achieve a collective and dependable workforce. Systems that are designed abroad for implementation in a different setting frequently get bogged down when imported into a developing environment that is not conducive for such systems. Differences in the perception of ethics are also an issue of concern in an outsourcing location. A lack of administrative vision and limited managerial capabilities limit the ability of outsource managers in developing countries. Both the physical and conceptual limitations on technical staff lead to administrative infeasibility that must be reckoned with. Overzealous entrepreneurs are apt to jump on opportunities to outsource production without a proper assessment of the capabilities of the receiving organization. Most often than not, outsourcing organizations do not fully understand the local limitations.

Some organizations take the risk of learning as they go, without adequate prior preparation.

TECHNICAL COMPATIBILITY

Hardware maintenance and software upgrade are, perhaps, the two most noticeable aspects of technical infeasibility of information technology in a developing country. The mistake is often made that once you install IT and all its initial components, you have the system for life. This is very far from the truth. The lack of proximity to the source of hardware and software enhancement makes this situation particularly distressing in a developing country. The technical capability of the personnel as well as the technical status of the hardware must be assessed in view of the local needs. Doing an overkill on the infusion of IT just for the sake of keeping up is as detrimental as doing nothing at all.

WORKFORCE INTEGRATION STRATEGIES

Any outsourcing enterprise requires adapting from one form of culture to another. The implementation of a new technology to replace an existing (or a nonexistent) technology can be approached through one of the several cultural adaptation options. Some suggestions are as follows:

Parallel interface: The host culture and the guest culture operate concurrently (side by side), with mutual respect on either side.

Adaptation interface: This is the case where either the host culture or the guest culture makes conscious effort to adapt to each other's ways. The adaptation often leads to new (but not necessarily enhanced) ways of thinking and acting.

Superimposition interface: The host culture is replaced (annihilated or relegated) by the guest culture. This implies cultural imposition on local practices and customs. Cultural incompatibility, for the purpose of business goals, is one reason to adopt this type of interface.

Phased interface: Modules of the guest culture are gradually introduced to the host culture over a period of time.

Segregated interface: The host and guest cultures are separated both conceptually and geographically. This used to work well in colonial days. But it has become more difficult with modern flexibility of movement and communication facilities.

Pilot interface: The guest culture is fully implemented on a pilot basis in a selected cultural setting in the host country. If the pilot implementation works with good results, it is then used to leverage further introduction to other localities.

HYBRIDIZATION OF INNOVATION CULTURES

The increased interface of cultures through industrial outsourcing is gradually leading to the emergence of hybrid cultures in many developing countries. A hybrid culture derives its influences from diverse factors, where there are differences in how the local population views education, professional loyalty, social alliances, leisure pursuits, and information management. A hybrid culture is, consequently, not fully embraced by either side of the cultural divide. This creates a big challenge to managing outsourcing projects.

Sustainability quality is at the intersection of efficiency, effectiveness, and productivity. Efficiency provides the framework for quality in terms of resources and inputs required to achieve the desired level of quality. Effectiveness comes into play with respect to the application of product quality to meet specific needs and requirements of an organization. Productivity is an essential factor in the pursuit of quality as it relates to the throughput of a production system. To achieve the desired levels of quality, efficiency, effectiveness, and productivity, a new research framework must be adopted for sustainability to take hold across national and cultural boundaries.

Several aspects of quality must undergo rigorous research along the realms of both quantitative and qualitative characteristics. Many times, quality is taken for granted and the flaws only come out during the implementation stage, which may be too late to rectify. The growing trend in product recalls is a symptom of *a-priori* analysis of the sources and implications of quality at the product conception stage. This column advocates the use of the DEJI systems model for enhancing quality design, quality evaluation, quality justification, and quality integration through hierarchical and stage-by-stage processes.

Better quality is achievable and there is always room for improvement in the quality of products and services. But we must commit more efforts to the research at the outset of the product development cycle. Even the human elements of the perception of quality can benefit from more directed research from a social and behavioral sciences point of view.

SUSTAINABILITY ACCOUNTABILITY

Throughout history, engineering has answered the call of the society to address specific challenges. With such answers comes a greater expectation of professional accountability. Consider the level of social responsibility that existed during the time of the Code of Hammurabi. Two of the laws are echoed below:

Hammurabi's Law 229:

If a builder builds a house for someone and does not construct it properly, and the house which he built falls in and kills its owner, then that builder shall be put to death.

Hammurabi's Law 230:

If it kills the son of the owner, the son of that builder shall be put to death.

These are drastic measures designed to curb professional dereliction of duty and enforce social responsibility with particular focus on product quality. Research and education must play bigger and more direct roles in the design, practice, and management of quality – present modern aspects of social responsibility in the context of day-to-day personal and professional activities. The global responsibility of the greater society is essential with respect to world development challenges covering the global economy, human development, global governance, and social relationships. Quality is the common theme in the development challenges. Focusing on the emerging field of Big Data, we should advocate engineering education collaboration, which aligns well with data-intensive product development. With the aforementioned principles as possible tenets for better research, education, and practice of quality in engineering and technology, this chapter suggests the DEJI model as a potential methodology. The model encourages the practice of building sustainability into a product right from the beginning so that the product integration stage can be more successful.

The design of quality in product development should be structured to follow point-to-point transformations. A good technique to accomplish this is the use of state-space transformation, with which we can track the evolution of a product from the concept stage to a final product stage. For the purpose of product quality design, the following definitions are applicable.

Product state: A state is a set of conditions that describes the product at a specified point in time. The *state* of a product refers to a performance characteristic of the product which relates input to output such that a knowledge of the input function over time and the state of the product at time $t=t_0$ determines the expected output for $t \geq t_0$. This is particularly important for assessing where the product stands in the context of new technological developments and the prevailing operating environment.

Product state-space: A product *state-space* is the set of all possible states of the product lifecycle. State-space representation can solve product design problems by moving from an initial state to another state, and eventually to the desired end-goal state. The movement from state to state is achieved by means of actions. A goal is a description of an intended state that has not yet been achieved. The process of solving a product problem involves finding a sequence of actions that represents a solution path from the initial state to the goal state. A state-space model consists of state variables that describe the prevailing condition of the product. The state variables are related to inputs by mathematical relationships. Examples of potential product state variables include schedule, output quality, cost, due date, resource, resource utilization, operational efficiency, productivity throughput, and technology alignment. For a product described by a system of components, the state-space representation can use state-by-state transfer of product characteristics. Each intermediate state may represent a significant milestone in the product. Thus, a descriptive state-space model facilitates an analysis of what actions to apply in order to achieve the next desired product state. A graphical representation can be developed for a product transformation from one state to another through the application of human or machine actions. This simple representation can be expanded to cover several components within the product information framework. Hierarchical linking of product elements provides an expanded transformation structure. The product state can be expanded in accordance with implicit requirements. These requirements might include grouping of design elements, linking precedence requirements (both technical and procedural), adapting to new technology developments, following required communication links, and accomplishing reporting requirements. The actions to be taken at each state depend on the prevailing product conditions. The nature of subsequent alternate states depends on what actions are implemented. Sometimes there are multiple paths that can lead to the desired end result. At other times, there exists only one unique path to the desired objective. In conventional practice, the characteristics of the future states can only be recognized after the fact, thus, making it impossible to develop adaptive plans. In the implementation of the

DEJI systems model, adaptive plans can be achieved because the events occurring inside and outside the product state boundaries can be taken into account.

If we describe a product by P state variables s_i, then the composite state of the product at any given time can be represented by a vector S containing P elements. The components of the state vector could represent either quantitative or qualitative variables (e.g., cost, energy, color, or time). We can visualize every state vector as a point in the state-space of the product. The representation is unique since every state vector corresponds to one and only one point in the state-space. Suppose we have a set of actions (transformation agents) that we can apply to the product information so as to change it from one state to another within the project state-space. The transformation will change a state vector into another state vector. A transformation may be a change in raw material or a change in design approach. The number of transformations available for a product characteristic may be finite or unlimited. We can construct trajectories that describe the potential states of a product evolution as we apply successive transformations with respect to technology forecasts. Each transformation may be repeated as many times as needed. Given an initial state S_0, the sequence of state vectors can be represented by successive state transitions.

EVALUATION OF SUSTAINABILITY QUALITY

A product can be evaluated on the basis of cost, quality, schedule, and meeting requirements. There are many quantitative metrics that can be used in evaluating a product at this stage. Learning curve productivity is one relevant technique that can be used because it offers an evaluation basis of a product with respect to the concept of growth and decay. The half-life extension (Badiru 2012) of the basic learning is directly applicable because the half-life of the technologies going into a product can be considered. In today's technology-based operations, retention of learning may be threatened by fast-paced shifts in operating requirements. Thus, it is of interest to evaluate the half-life properties of new technologies as the impact on the overall product quality. Information about the half-life can tell us something about the sustainability of learning-induced technology performance. This is particularly useful for designing products whose lifecycles stretch into the future in a high-tech environment.

JUSTIFICATION OF SUSTAINABILITY QUALITY

We need to justify an innovation program on the basis of quantitative value assessment. The systems value model is a good quantitative technique that can be used for innovation justification on the basis of value. The model provides a heuristic decision aid for comparing project alternatives. It is presented here again for the present context. Value is represented as a deterministic vector function that indicates the value of tangible and intangible attributes that characterize the product. It can be represented as V, which is the assessed value based on the contributing attributes of the components making up the product. Examples of product attributes are quality, throughput, manufacturability, capability, modularity, reliability, interchangeability, efficiency, and cost performance. Attributes are considered to be a combined function of factors. Examples of product factors are market share, flexibility, user acceptance, capacity utilization, safety, and design functionality. Factors are themselves considered to be composed of indicators. Examples of indicators are debt ratio, acquisition volume, product responsiveness, substitutability, lead time, learning curve, and scrap volume. By combining the aforementioned definitions, a composite measure of the operational value of a product can be quantitatively assessed. In addition to the quantifiable factors, attributes, and indicators that impinge upon overall project value, the human-based subtle factors should also be included in assessing overall project value.

EARNED VALUE TECHNIQUE FOR INNOVATION

Value is synonymous with quality. Thus, the contemporary earned value technique is relevant for "earned quality" analysis. This is a good analytical technique to use for the justification stage of the DEJI systems model. This will impact cost, quality, and schedule elements of product development with respect to value creation. The technique involves developing important diagnostic values for each schedule activity, work package, or control element. The variables are as follows: PV: planned value; EV: earned value; AC: actual cost; CV: cost variance; SV: schedule variance; EAC: estimate at completion; BAC: budget at completion; and ETC: estimate to complete. This analogical relationship is a variable research topic for quality engineering and technology applications.

INTEGRATION OF SUSTAINABILITY QUALITY

Without being integrated, a system will be in isolation and it may be worthless. We must integrate all the elements of a system on the basis of alignment of functional goals. The overlap of systems for integration purposes can conceptually be viewed as projection integrals by considering areas bounded by the common elements of subsystems. Quantitative metrics can be applied at this stage for effective assessment of the product state. Trade-off analysis is essential in quality integration. Pertinent questions include the following:

> What level of trade-offs on the level of quality is tolerable?
> What is the incremental cost of higher quality?
> What is the marginal value of higher quality?
> What is the adverse impact of a decrease in quality?
> What is the integration of quality of time?

The guidelines and important questions relevant for quality integration are as follows:

- What are the unique characteristics of each component in the integrated system?
- How do the characteristics complement one another?
- What physical interfaces exist among the components?
- What data/information interfaces exist among the components?
- What ideological differences exist among the components?
- What are the dataflow requirements for the components?
- What internal and external factors are expected to influence the integrated system?
- What are the relative priorities assigned to each component of the integrated system?
- What are the strengths and weaknesses of the integrated system?
- What resources are needed to keep the integrated system operating satisfactorily?
- Which organizational unit has primary responsibility for the integrated system?

The proposed approach of the DEJI system model will facilitate a better alignment of product technology with future development and needs. The stages of the model require research for each new product with respect to design, evaluation, justification, and integration. Existing analytical tools and techniques can be used at each stage of the model.

UMBRELLA THEORY FOR INNOVATION

Extensive literature review concludes that an overarching theory was lacking to guide the process of innovation. The key to a successful actualization of innovation centers on how people work and behave in team collaborations. Hence, the proposed methodology of "umbrella theory for innovation" takes into account the interplay between people, tools, and process. The umbrella theory for innovation capitalizes on the trifecta of human factors, process design, and technology tool availability within the innovation environment. The theory harnesses the proven efficacies of existing tools and principles of systems engineering and management. Two specific techniques in this regard are the Triple C model and the DEJI systems model.

A semantic network, also called a frame network, is a knowledge base that represents semantic relationships between elements in an operational network or system. It is often used for knowledge representation purposes in software systems. In innovation, a semantic network can be used to represent the relationships among elements (people, technology, and process) in the innovation system. This representation can give a visual cue of the critical paths in the innovation network. The requirements for the success of an innovation effort include the following:

1. **Relative advantage**: This is the degree to which an innovation is perceived as better than the idea it supersedes by a particular group of users, measured in terms that matter to those users, such as economic advantage, social prestige, convenience, or satisfaction. The greater the perceived relative advantage of an innovation, the more rapid its rate of adoption is likely to be. There are no absolute rules for what constitutes "relative advantage." It depends on the particular perceptions and needs of the user group.
2. **Compatibility with existing values and practices**: This is the degree to which an innovation is perceived as being consistent with the values, past experiences, and needs of potential adopters.

An idea that is incompatible with their values, norms, or practices will not be adopted as rapidly as an innovation that is compatible.

3. **Simplicity and ease of use**: This is the degree to which an innovation is perceived as difficult to understand and use. New ideas that are simpler to understand are adopted more rapidly than innovations that require the adopter to develop new skills and understandings.

4. **Trialability**: This is the degree to which an innovation can be experimented with on a limited basis. An innovation that is triable represents less uncertainty to the individual who is considering it.

5. **Observable results**: The easier it is for individuals to see the results of an innovation, the more likely they are to adopt it. Visible results lower uncertainty and also stimulate peer discussion of a new idea, as friends and neighbors of an adopter often request information about it.

INNOVATION READINESS MEASURE

Badiru (2019) presented the framework for an innovation assessment tool. The tool is designed to assess the readiness of an organization on the basis of desired requirements with respect to pertinent factors.

Based on the spread of innovation requirements over the relevant factors, a quantitative measure of the innovation readiness of the organization can be formulated as follows:

Assuming that each checkmark can be rated on a scale of 0 to 10, the following composite measure can be derived:

$$IR = \sum_{i=1}^{N} \sum_{j=1}^{M} r_{ij},$$

where:

IR = innovation readiness of the organization;

N = number of requirements;

M = number of factors; and

r_{ij} = alignment measure of requirement i with respect to factor j.

The aforementioned measure can be normalized on a scale of 0 to 100 on the basis of which organizations and/or units within an organization can be compared and assessed for innovation readiness. Obviously, an organization that is competent in executing and actualizing innovation will yield a higher innovation readiness measure.

INNOVATION RISK MANAGEMENT

Risk is an essential element of innovation. No risk means no accomplishment. The important thing is to manage risk constructively. Risk management is an integral part of innovation. For innovation, particularly for those dealing with new ventures, risk management can be carried out effectively by investigating and identifying the sources of risks associated with each activity. These risks can be assessed or measured in terms of likelihood and impact. Because of the exploration basis of new technology, a different and diverse set of risk concerns will be involved. So as risks are assessed for managerial processes, technical and managerial risks must also be assessed. The major activities in innovation analysis consist of feasibility studies, design, transportation, utility, survey works, construction, permanent structure works, mechanical and electrical installations, maintenance, and so on.

Definition of Risk

Risk is often ambiguously defined as a measure of the probability, level of severity, and exposure to all hazards for a project activity. Practitioners and researchers often debate the exact definition, meaning, and implications of risk. Two alternate definitions of risk are presented next:

Risk is an uncertain event or condition that, if it occurs, has a positive or negative effect on a project objective.

Risk is an uncertain event or set of circumstances that, should it occur, will have an effect on the achievement of the project's objectives.

In this book, we present the following definition of risk management:

Risk management is the state of having a contingency ready to respond to the impact (good or bad) of occurrence of risk, such that risk mitigation or risk exploitation becomes an intrinsic part of the project plan.

For any innovation undertaking, there is always a chance that things will not turn out exactly as planned. Thus, project risk pertains to the probability of uncertainties of the technical, schedule, and cost outcomes of the project. All technology-based projects are complex and they involve risks in all the phases of the project starting from the feasibility phase to the operational phase. These risks have a direct impact on the project schedule, cost, and performance. These projects are inherently complex and volatile with many variables. A proper risk mitigation plan, if developed for identified risks, would ensure better and smoother achievement of project goals within the specified time, cost, and technical requirements. Conventional project management techniques, without a risk management component, are not sufficient to ensure time, cost, and quality achievement of a large-scale project, which may be mainly due to changes in scope and design, changes in government policies and regulations, changes in industry agreement, unforeseen inflation, underestimation, and improper estimation. Projects that are exposed to such risks and uncertainty can be effectively managed with the incorporation of risk management throughout the projects' lifecycle.

Sources of Uncertainty

Project risks originate from the uncertainty that is present in all projects to one extent or another. A common area of uncertainty is the size of project parameters, such as time, cost, and quality with respect to the expectations of the project. For example, we may not know precisely how much time and effort will be required to complete a particular task. Possible sources of uncertainty include the following:

Poor estimates of time and cost
Lack of a clear specification of project requirements
Ambiguous guidelines about managerial processes
Lack of knowledge of the number and types of factors influencing the project
Lack of knowledge about the interdependencies among activities in the project
Unknown events within the project environment
Variability in project design and logistics
Project scope changes
Varying direction of objectives and priorities
Impacts of regulations

Risks can be mitigated, not eliminated. In fact, risk is the essence of any enterprise. In spite of government regulations designed to reduce accident risks, accidents will occasionally happen. Government regulators can work with organizations to monitor data and operations. This will only preempt a fraction of potential risks of incidents. For this reason, government agencies must work with organizations to ensure that adequate precautions are taken in all operating scenarios. Government and industry must work together in a risk-mitigation partnership, rather than in an adversarial and dictatorial relationship. There is no risk-free activity in business and industry of today. For example, many of the safety and security incidents observed over the years involved human elements – errors, incompetence, negligence, and so on. How do you prevent negligence? You can encourage non-negligent operation or incentivize perfect record, but humans will still be humans when bad things happen. Effective risk management requires a reliable risk analysis technique. The following points are how to deal with risk management:

Avoid
Assign
Assume
Mitigate
Manage

A four-step process of managing risk is as follows:

STEP ONE – Identify the risks
STEP TWO – Assess the risks
STEP THREE – Plan risk mitigation
STEP FOUR – Communicate risk

We must venture out on the risk limb in order to benefit from what the innovation offers. Many leaders profess the call of "taking risk," but guidance is often lacking on to what extent risks can be taken. A quote that typifies the benefit of taking risk is echoed here:
Consider the following quote:

"Behold the lowly Turtle – he only makes

progress when he sticks his neck out."

— James Conan Bryan)

Let us take another look at the basic definition of risk:

Risk – "Potential Realization of an Unwanted Negative Consequence"
Reward – "Potential Realization of a Desired Positive Consequence"

A master list of risk management involves the following:

• New technology
• Functional complexity
• New versus replacement
• Leverage on company
• Intensity of business need
• Interface existing applications
• Staff availability

- Commitment of team
- Team morale
- Applications knowledge
- Client information system (IS) knowledge
- Technical skills availability
- Staff conflicts
- Quality of information available
- Dependability on other projects
- Conversion difficulty
- End-date dictate
- Conflict resolution mechanism
- Continued budget availability
- Project standards used
- Large/small project
- Size of team
- Geographical dispersion
- Reliability of personnel
- Availability of support organization
- Availability of champion
- Vulnerability to change
- Stability of business area
- Organizational impact
- Tight time frame
- Turnover of key people
- Change budget accepted
- Change process accepted
- Level of client commitment
- Client attitude toward IS
- Readiness for takeover
- Client design participation
- Client participation in acceptance test
- Client proximity to IS
- Acceptance process

Possible risk response planning can follow the following options:

> Accept – Do nothing because the cost to fix is more expensive than the expected loss
> Avoid – Elect not to do part of the project associated with the risk
> Contingency planning – Frame plans to deal with risk consequence and monitor risk regularly (identify trigger points)
> Mitigate – Reduce either the probability of occurrence, the loss, or both
> Transfer – Outsource

The perspectives and guidance offered by the umbrella model for innovation management can create an avenue for managing, controlling, or mitigating risk in innovation pursuits.

CONCLUSION

It is expected that the methodology of the umbrella theory of innovation will inspire new research inquiries into the human factors of driving innovation in organizations. The theory has the benefit of providing coverage for all the typical nuances (qualitative and quantitative) that may be encountered in the innovation environment. Of particular importance is the consideration of the people factors of innovation. The common flawed view of innovation is that it is predicated on the acquisition of technological items. While technology may be the underpinning of a specific innovation project, more often than not, the human factors will determine the success or failure of innovation. The umbrella theory explicitly calls out the people aspects of the pursuit of innovation. The quantitative measure of innovation readiness can be adapted and expanded to fit specific research, development, and implementation themes related to the pursuit of innovation. Quality is an integrative process that must be evaluated on a stage-by-stage approach. This requires research, education, and implementation strategies that consider several pertinent factors. This chapter suggests the DEJI systems model, which has been used successfully for product development applications, as a viable methodology for quality design, quality evaluation, quality justification, and quality integration. This chapter is intended as a source to spark the interest of researchers to apply this tool in new product development efforts.

REFERENCES

Badiru, Adedeji B. (Fall 2012), "Application of the DEJI Model for Aerospace Product Integration," *JAAP*, Vol. 2, No. 2, pp. 20–34.
Badiru, Adedeji B. (2019), *Systems Engineering Models: Theory, Methods, and Applications*, Taylor & Francis/CRC Press, Boca Raton, FL.

Education and Sustainability

5

"Education misapplied is education missed."

— Adedeji Badiru

INTRODUCTION

Education is the most powerful tool that can be used toward sustainability. Education personally, economically, or socially is important to increasing the potential for achieving the goals of sustainability. Education, through explicit communication, is essential for increasing awareness of sustainability requirements. The quote by George Bernard Shaw, an Irish playwright, reminds us that "The single biggest problem in communication is the illusion that it has taken place." Sustainability practices save money through awareness and communication, while reducing environmental predicaments. Engagement of individuals whether personally or professionally will help engage people to understand their direct responsibility and the effects they have on the environment. When businesses are going through "tough times," the number one thing to do is cut costs. Many think cutting jobs is the first answer, but they should think first about how to be more sustainable. Innovation mixed with business processes can change the mindsets of people, businesses, and reduce incremental costs. Education plays a vital role with these processes because it changes people's ways of thinking. y . The methodology is not only for certain tasks that they perform while at work, but also to affect activities in their everyday lives. This is a life change, not a flavor of the month. Without making the changes and investing in education, our entire environment will suffer consequences (Agustiady and Badiru 2013; Badiru 2010, 2015, 2019).

Education can be the easiest part of sustainability. Education includes engagement, motivation, teaching, and changes of everyday processes all leading to continual improvements toward our day-to-day lives. Forming teams for sustainability is a great initiative to begin educating others. As a team, each person puts forward his or her own understanding and education of sustainability and has a collaborative effort toward the initiative. The beginning process of this is to form a proper team. The team must be cross-functional, knowing vast areas of the entire environment. A project charter is the next step for this team so that there is an executive summary and understanding of what is to be accomplished. Once the project charter is developed, a timeline is imperative so that the focus does not diminish and the priority remains high. Senior management should be given a presentation on the group's objectives and goals along with permission for resources for the effort. A definition for sustainability should be identified next. This definition can vary slightly, but should include some of the following terms or ideas:

> Sustainability is the human preservation of the environment, whether personally, socially, or economically through responsible management of resources, education and continuous process improvement. Being sustainable corporate citizens will increase sales and profitability by reducing costs and having a competitive advantage.

Once it is defined on what sustainability means, the next steps are the action steps. Communication is a must on what sustainability means to the remaining core groups, which is then cascaded down to everyone involved in the corporation or area. An action plan should come next on how to be sustainable and the steps involved within. Each business is different, but commonalities are energy reduction, utilizing automated thermostats, turning off computers, utilizing less water, and decreasing gas hot water heaters.

The following statement was written by the participants in Business Sector Team calls of the US Partnership for Education for Sustainable Development:

> "To meet the immense challenges of the present and the future, it is important that all undergraduate and graduate college students learn about our environmental and social sustainability challenges and be provided with learning opportunities that engage them in solutions to these challenges. We live in a unique time, where the decisions of this generation may very well dictate the health of the planet for this and future generations. The impacts of these decisions will affect the quality of life across the globe. All students need to learn, through an interdisciplinary approach, not only the specifics of our sustainability challenges and the possible solutions, but also the interpersonal skills, the systems thinking skills, and the change agent skills to effectively help to create a more sustainable future. We are looking for these sustainability-educated students as future business people, as employees, as

consumers, innovators, government leaders and investors. We would like to see this be a requirement for all students."

Stimulus money is increasingly rising for education toward green jobs. The federal economic stimulus plan allocated more than $70 billion in direct spending, tax breaks, and loan guarantees for the energy in the nation, mostly the "green" energy.

SUSTAINABILITY EDUCATION

Whoever owns the knowledge controls the power. Education should play a vital role in sustainability programs. In these days of knowledge economy, adequate education is needed to succeed in any workplace whether it is a hamburger stand, an administrative office, or a manufacturing plant. Even in some industrialized nations, a large percentage of the adult population in neglected communities is functionally illiterate and less likely to embrace sustainability. It used to be that the children of these poverty-stricken communities were needed to drive the wheels of manual labor in local factories. But modern industries, with increasing push for automation, do not need much of the services of the low-grade workers. Knowledge workers are the norm in the present economy. Education geared toward the new sustainability direction will, thus, be needed to participate actively in the society. Poor parents always proclaim that they do not want their children to end up where they did. Yet only limited educational efforts are made to ensure that they do not. The facilitation of sustainability can follow three paths:

- Communication of the benefits and threats to sustainability
- Knowledge transfer about sustainability
- Information sharing about opportunities to advance sustainability programs

University–Industry Sustainability Partnership

Knowledge is an everlasting capital. The establishment of a formal process for the interface of institutions of higher learning and industry can be one of the capitals for achieving sustainability. Universities have unique capabilities that can be aligned with industry capabilities to produce symbiotic working relationships. Private industrial research projects must complement public industrial research ventures for the sake of advancing sustainability.

Academic institutions have a unique capability to generate, learn, and transfer technology to industry. The quest for knowledge in academia can fuel the search for innovative solutions to specific sustainability problems. Cooperating industry is a fertile ground for developing prototypes of new innovation. Industrial settings are good avenues for practical implementation of technology. Industry-based implementation of university-developed sustainability technology can serve as the impetus for further efforts to develop new technology. Technologies that are developed within the academic community mainly for research purposes often fade into oblivion because of the lack of formal and coordinated mechanism for practical implementation. The potentials of these technologies go untapped for several reasons including the following:

- The developer does not know which industry may need the technology.
- Industry is not aware of the technology available in academic institutions.
- There is no coordinated mechanism for technical interface between industry and university groups.

Universities interested in technology development are often hampered by the lack of adequate resources for research and training activities. Industry can help in this regard by providing direct support for industrial groups to address specific industrial problems related to sustainability. The universities also need real problems to work on as projects or case studies. Industry can provide these under a cooperative arrangement.

The respective needs and capabilities of universities and industrial establishments can be integrated symbiotically to provide benefits for each group. University courses offered at convenient times for industry employees can create opportunity for university–industry interaction. Class projects for industry employees can be designed to address real-life sustainability problems. This will help industry employees to have focused and rewarding projects. Class projects developed in the academic environment can be successfully implemented in actual work environments to provide tangible benefits. With a mutually cooperative interaction, new sustainability developments in industry can be brought to the attention of academia while new academic research developments can be tested in industrial settings.

Sustainability Clearinghouse

Academic institutions can serve as convenient locations for technology clearinghouses. Such clearinghouses can be organized to provide up-to-date information for sustainability activities. Specific industrial problems can be studied at the clearinghouse. The clearinghouse can serve as a repository for information on various technology tools for sustainability. Industry would participate in the clearinghouse through the donation of equipment, funds, and personnel time. The services provided by a clearinghouse could include a combination of the following:

- Provide consulting services on technology to industry
- Conduct on-site short courses with practical projects for industry
- Serve as a technology library for general information
- Facilitate technology transfer by helping industry identify which technology is appropriate for which problems
- Provide technology management guidelines that will enable industry to successfully implement new technology in existing operations
- Expand training opportunities for engineering students and working engineers

Center of Excellence for Sustainability

As the interest in sustainability spreads, there will be a need to differentiate capabilities among the various players. In its oversight roles, the government can sponsor cooperative interactions between academia and industry by providing broad-based funding mechanisms. As an example, the US National Science Foundation has a program to provide funds for Industry/University Cooperative Research Centers. This sort of partnership can be leveraged for the pursuit of sustainability programs. The leaders of a nation pursuing sustainability should actively support cooperative efforts between universities and industry. The establishment of centers of excellence for pursuing sustainability-related research is one approach to creating an atmosphere that is conducive for industry–university interaction. Several states in the United States have used this approach to address specific development needs.

THE ROLE OF WOMEN IN ACHIEVING SUSTAINABILITY

Even though women may not generally be the head of the household, they are, nonetheless, leaders of the home. In this leadership role, women can be key players in ensuring sustainability, regardless of whichever sustainability definition is in operation. Women play important roles in national development and should lead societal efforts in sustainability. Women have significant capabilities in terms of political, economic, and industrial roles in a nation. In developing nations particularly, women have played important economic roles as traders for a long time. The economic power garnered from their entrepreneurial activities should be transformed to decision-making power to support sustainability programs. In very democratic nations, the political power of women has been more pronounced primarily through voting rights. These rights should be directed at policy-making endeavors that can facilitate sustainability. Women groups have emerged in some developing nations to better utilize their collective powers. Many of the efforts of these groups have been directed at development. For example, the National Council for Women Development has been very active politically in Ghana. In Nigeria, the Rural Women Development program, spearheaded by the first lady of the nation, is directed at improving the economic impact of rural women. Small-scale industries owned by women have been one of the tangible outputs of these women movements. Other examples abound in developed and developing nations around the world. It is the view of the authors that there should be a global coalition of women for sustainability, whereby an educational platform will be a primary strategy.

Agriculture and Sustainability

Agriculture is one avenue through which the impacts and benefits of sustainability can be readily noticed. A hungry society cannot be an industrially productive society and cannot be focused on sustainability initiatives. Feed the masses to be content, then they will have room to consider and embrace sustainability. It is generally believed that an underdeveloped economy is characterized by an agricultural base. Based on this erroneous belief, several developing nations have abandoned their previously solid agricultural base in favor of "unsustainable" industrialization. The fact is that a strong agricultural base is needed to complement industrialization programs. Agriculture, itself, is a good target for modernization and industrialization in ways that complement and support sustainability goals. If industrialization does not yield immediate benefits, the society will be exposed to the double jeopardy of hunger and material deprivation. Once abandoned, agriculture is a difficult process to recoup and nonsustainable practices will creep in. Since agricultural processes take several decades to perfect, revitalization of abandoned agriculture may require several decades. Agriculture should play a major role in the foundation for sustainable industrial development. The agricultural sector can serve as a viable market for a developed industry.

Evolution of Efficient Agriculture

It is interesting to note how the Agricultural Revolution led to the Industrial Revolution in the past. Human history indicates that humans started out as nomad hunters and gatherers, drifting to wherever food could be found. About 12,000 years ago, humans learned to domesticate both plants and animals. This agricultural breakthrough allowed humans to become settlers, thereby spending less time in search of food. More time was, thus, available for pursuing innovative activities, which led to discoveries of better ways of planting and raising animals for food. That initial agricultural discovery eventually paved the way for the Agricultural Revolution. During the Agricultural Revolution, mechanical devices, techniques, and storage mechanisms were developed to aid the process of agriculture. These inventions made it possible for more food to be produced by fewer people. The abundance of food meant that more members of the community could spend that time for other pursuits rather than the customary labor-intensive agriculture. Naturally, these other pursuits involved the development and improvement of the tools of agriculture. The extra free time brought on by more efficient agriculture was, thus, used to bring about minor technological improvements in agricultural implements. These more advanced agricultural tools led to even more efficient agriculture. The transformation from the digging stick to the metal hoe is a good example of the raw technological innovation of that time. Sustainability was possible then, and it should be possible now.

Emergence of Cities

With each technological advance, less time was required for agriculture, thereby, permitting more time for further technological advancements. The advancements in agriculture slowly led to more stable settlement patterns. These patterns led to the emergence of towns and cities. With central settlements away from farmlands, there developed a need for transforming agricultural technology to domicile technology that would support the new city life. The transformed technology was later turned to other productive uses, which eventually led to the emergence of the Industrial Revolution. To this day, the intertwined relationships between agriculture and industry can still be seen, although one would have to look harder and closer to see them from the standpoint of sustainability.

Human Resources for Sustainability

Technical, administrative, and service manpower will be needed to support sustainability programs. People make development possible. People must sustain sustainability activities. No matter how technically capable a machine may be, people will still be required to operate or maintain it. Soon after World War II, it was generally believed that physical capital formation was a sufficient basis for development. That view was probably justified at that time because of the role that machinery played during the war. It was not obvious then that machines without a trained and skillful work force did not constitute a solid basis for development. It has now been realized that human capital is as crucial to development and sustainability as physical capital is. The investment in human resource development through education must be given a high priority in the overall sustainability strategy. Some of the important aspects of manpower supply analysis for industrial development include:

- Level of skills required
- Mobility of the manpower
- The nature and type of skills required
- Retention strategies to reduce brain drain
- Potential for coexistence of people and technology
- Continuing education to facilitate adaptability to technology changes

The Role of Technology in Sustainability

Technology can facilitate sustainability. But technology must be managed properly to play an effective role. There is a multitude of new technologies that has emerged in recent years. Hardware and software technologies are playing increasingly more roles in sustainability programs. But much more remains to be done in educational programs to develop new technologies specifically for sustainability. It is important to consider the peculiar characteristics of a new technology before establishing adoption and implementation strategies for sustainability. The justification for the adoption of a new technology should be a combination of several factors rather than a single characteristic of the technology. The important characteristics to consider include productivity improvement, improved product quality, reduction in production cost, flexibility, reliability, and safety. An integrated evaluation must be performed to ensure that a proposed technology is justified both economically and technically. The scope and goals of the proposed technology must be established right from the beginning of a sustainability project. This entails the comparison of industry objectives with the overall national goals in the areas discussed next.

Market target: This should identify the customers of the proposed technology. It should also address items such as market cost of the proposed product, assessment of competition, and market share.

Growth potential: This should address short-range expectations, long-range expectations, future competitiveness, future capability, and prevailing size and strength of the competition.

Contributions to sustainability goals: Any prospective technology must be evaluated in terms of direct and indirect benefits to be generated by the technology. These may include product price versus value, increase in international trade, improved standard of living, cleaner environment, safer workplace, and improved productivity.

Profitability: An analysis of how the technology will contribute to profitability should consider past performance of the technology, incremental benefits of the new technology versus conventional technology, and value added by the new technology.

Capital investment: Comprehensive economic analysis should play a significant role in the technology assessment process. This may cover an evaluation of fixed and sunk costs, cost of obsolescence, maintenance requirements, recurring costs, installation cost, space requirement cost, capital substitution potentials, return on investment, tax implications, cost of capital, and other concurrent projects.

Skill and resource requirements: The utilization of resources (manpower and equipment) in the pre-technology and post-technology phases of industrialization should be assessed. This may be based on material input/output flows, high value of equipment versus productivity improvement, required inputs for the technology, expected output of the technology, and utilization of technical and nontechnical personnel.

Risk exposure: Uncertainty is a reality in technology adoption efforts. Uncertainty will need to be assessed for the initial investment, return on investment, payback period, public reactions, environmental impact, and volatility of the technology.

National sustainability improvement: An analysis of how the technology may contribute to national sustainability goals may be verified by studying industry throughput, efficiency of production processes, utilization of raw materials, equipment maintenance, absenteeism, learning rate, and design-to-production cycle.

DEJI Model for Sustainability Assessment

In terms of assessing technology for sustainability implementation, one approach that can be used is the Design, Evaluate, Justify, and Integrate (DEJI) systems model. The model is unique among process improvement tools and techniques because it explicitly calls for a justification of the technology within the process improvement cycle. This is important for the purpose of determining when a technology should be terminated even after going into production. If the program is justified, it must then be integrated and "accepted" within the ongoing sustainability program of the enterprise. The DEJI systems model can be applied across the spectrum of the following elements of an organization:

1. People
2. Process
3. Technology

Foundation for Sustaining Sustainability

Sustainability built upon a solid foundation can hardly fail. The major ingredients for durable sustainable industrial, economic, and technological developments are electrical power, water, transportation, and communication facilities. These items should have priority in major sustainability programs:

Primary Amenities:

- Reliable power supply
- Consistent water supply
- Good transportation system
- Efficient communication system

Supporting Amenities:

- Housing
- Education
- Health Care

CONCLUSIONS

The provision of adequate healthcare facilities is particularly essential for building a strong industrial base and ensuring sustainability. A healthy society is a productive society; a sick society will be a destitute society and more susceptible to infringing on sustainability. Diseases that often ravage impoverished nations can curtail the productive capabilities of the citizens. The destructive effects of many of these diseases can be stemmed by prompt access to basic healthcare services. The worldwide emergence of COVID-19 in 2020 has highlighted the need for urgency in this regard. Education for sustainability is the collective responsibility of everyone.

REFERENCES

Agustiady, Tina and Badiru, Adedeji B. (2013), *Sustainability: Utilizing Lean Six Sigma Techniques*, Taylor & Francis CRC Press, Boca Raton, FL.

Badiru, Adedeji B. (Nov. 2010), "The Many Languages of Sustainability," *Industrial Engineer*, Vol. 42, No. 11, pp. 31–34.

Badiru, Adedeji B. (2015), "A Systems Model for Global Engineering Education: The 15 Grand Challenges," *Engineering Education Letters*, Vol. 1, No. 1, pp. 31–34, Open Access Link: 2015:3 http://dx.doi.org/10.5339/eel.2015.3.

Badiru, Adedeji B. (Jan. 2019), "Our Greatest Grand Challenge: To Address Society's Urgent Problems, Engineers Need to Step Up to the Political Plate," *ASEE PRISM*, Vol. 28, No. 1, pp. 56.

Lean and Waste Reduction

6

"Haste makes waste, just as rush makes ruin in sustainability."

– Adedeji Badiru

INTRODUCTION: LEAN THINKING

Kilchiro Toyoda, founder of Toyota, said, "Every defect is a treasure, if the company can uncover its cause and work to prevent it across the corporation." The lesson that is conveyed by this insightful quote is that the elimination of waste in the production system can translate to a valuable treasure for the organization. Thus, the goal is to find and eliminate waste in the pursuit of whatever our goals are. This concept is directly applicable to the pursuit of sustainability.

"Lean" is a terminology that is well known and defined as an elimination of waste in operations through managerial principles. Many principles are comprised in the Lean concept, but the major thought to remember is effective utilization of resources and time in order to achieve higher-quality products and ensure customer satisfaction. Remembering back, defects are anything that the customer is unhappy with and is a term utilized in Six Sigma. Six Sigma identifies and eliminates these defects so that the customer in turn is satisfied. The customer is the number one focus and if they are unhappy, they will have no problem going elsewhere, which most likely is a competition for the business. Coupling Lean and Six Sigma will reduce waste and reduce defects. The concept will be called Lean Six Sigma going further.

The most basic concept when discussing waste reduction begins with Kaizen. Kaizen is a Japanese concept defined as "taking apart and making better." The concept takes a vast amount of project management techniques to facilitate the process going forward. 5s processes are the most predominant and commonly known for Kaizen events.

The 5s principles are determined by finding a place for everything and everything in its place.

The 5s levels are as follows:

Sort – Identify and eliminate necessary items and dispose of unneeded materials that do not belong in an area. This reduces waste, creates a safer work area, opens space, and helps visualize processes. It is important to sort through the entire area. The removal of items should be discussed with all personnel involved. Items that cannot be removed immediately should be tagged for subsequent removal.

Sweep – Clean the area so that it looks like new and clean it continuously. Sweeping prevents an area from getting dirty in the first place and eliminates further cleaning. A clean workplace indicates high standards of quality and good process controls. Sweeping should eliminate dirt, build pride in work areas, and build value in equipment.

Straighten – Have a place for everything and everything in its place. Arranging all necessary items is the first step. It shows what items are required and what items are not in place. Straightening aids efficiency; items can be found more quickly and employees travel shorter distances. Items that are used together should be kept together. Labels, floor markings, signs, tape, and shadowed outlines can be used to identify materials. Shared items can be kept at a central location to eliminate purchasing more than needed.

Schedule – Assign responsibilities and due dates to actions. Scheduling guides sorting, sweeping, and straightening and prevents regressing to unclean or disorganized conditions. Items are returned where they belong and routine cleaning eliminates the need for special cleaning projects. Scheduling requires checklists and schedules to maintain and improve neatness.

Sustain – Establish ways to ensure maintenance of manufacturing or process improvements. Sustaining maintains discipline. Utilizing proper processes will eventually become routine. Training is the key to sustaining the effort and involvement of all parties. Management must mandate the commitment to housekeeping for this process to be successful.

The benefits of 5s include (1) a cleaner and safer workplace; (2) customer satisfaction through better organization; and (3) increased quality, productivity, and effectiveness.

Kai is defined as to break apart or disassemble so that one can begin to understand. Zen is defined as to improve. This process focuses on improvements objectively by breaking down the processes in a clearly defined and understood manner so that wastes are identified, improvement ideas are created, and wastes are both identified and eliminated. The philosophy includes reducing cycle times and lead times in turn increasing productivity, reducing work-in-process (WIP), reducing defects, increasing capacity, increasing flexibility, and improving layouts through visual management techniques.

Operator cycle times need to be understood in order to reduce the nonproductive times. Operators should also be cross-functional so that they are able to perform different job functions and the workloads of each function are well balanced. The work performed needs to be not only value-added work, but also work that is in demand through customers. WIP should be eliminated to reduce inventory. Inventory should be seen simply as money waiting in process and should be reduced as much as possible. WIP can be reduced by reducing set-up times, transporting smaller quantities of batch outputs, and line balancing. Bottlenecks should be removed by finding nonvalue-added tasks and removing the excess time spent by both machinery and humans. Flexible layouts promote efficiency in the 5Ms, which are defined next.

Sometimes an eighth waste is added in and an abbreviation of DOWN-TIME is associated with the acronym. It is defined as follows:

- Defect/correction
- Overproduction
- Waiting
- Not utilizing employee talents
- Transportation/material movement
- Inventory
- Motion
- Excessive processing

The primary technique for reducing waste and defects utilizing Lean Six Sigma techniques is demonstrated next in order of operations:

ROOT CAUSE ANALYSIS

1. Process mapping
2. Data gathering – gather data on any process with defects and issues. Utilize voice of the customer to find what data are needed
3. Cause/effect analysis (seeking root cause) utilizing 5Ms
4. Verifying root cause with data-driven results – ask why five times to ensure whether the proper root cause is found. Do not Band-Aid problems, instead eliminate the cause they are occurring

5. Solutions and continuous improvement plans
6. Test implementation plan by piloting. Pilot plans can include actual trials or mock trials with details information
7. Implement continuous improvement ideas
8. Control/monitoring plan
9. Documentation of lessons learned

Now that the process is laid out in terms of making proper improvements, the sustainability portion must be realized. The control plan during this phase is crucial. The control plan must not just be documented, but it should be a living document that is followed structurally. The accountability portion for this phase is a key portion in having lasting results. The more specific the plan is, the better off the implementation of it will be. The plans must also be attainable or the plan will fail.

Lean can also involve some statistical tools. The tools demonstrate the efficiencies and labor balancing. The main statistical tools are as follows:

First pass yield (FPY) indicates the number of good outputs from a first pass at a process or step. The formula is as follows:

$$FPY = (\# \text{ accepted})/(\# \text{ processed})$$

The formula for the FPY ratio is % FPY=[(# accepted)/(# processed)] × 100.

This number does not include reworked product that was previously rejected.

Rolled throughput yield (RTY) covers an entire process. If a process involves three activities with FPYs of 0.90, 0.94, and 0.97, the RTY would be 0.90 × 0.94 × 0.97=0.82. The %RTY=0.82 × 100=82%.

Value-added time (VAT) is % VAT=(sum of activity times)/(lead time) × 100. When the sum of activity times equals lead time, the VAT is 100%. For most processes, % VAT=5% to 25%. If the sum of activity times equals the lead time, the time value is not acceptable and activity times should be reduced.

Takt time is a Kaizen tool used in the order-taking phase. *Takt* is a German word for pace. Takt time is defined as time per unit. This is the operational measurement to keep production on track. To calculate takt time, the formula is time available/production required. Thus, if a required production is 100 units per day and 240 minutes are available, the takt time = 240/100 or 2.4 minutes to keep the process on track. Individual cycle times should be balanced to the pace of takt time. To determine the number of employees required, the formula is (labor time/unit)/takt time. Takt in this case is time per unit. Takt requires visual controls and helps reduce accidents and injuries in the workplace. Monitoring inventory and production WIP will reduce waste or *muda*. *Muda* is a Japanese term for waste where waste is defined as any activity that consumes some type of resource but is nonvalue added for the customer. The customer is not willing to pay for this resource because it is not benefiting them. Types of *muda* include scrap, rework, defects, mistakes, and excess transport, handling, or movement.

The Lean house is a common methodology for understanding Lean and waste reduction.

Mistake proofing is a subject of its own when brought into the Lean Six Sigma methodology. This term is often called Poka Yoke, also known as another initiative to improve production systems. The methodology eliminates product defects before they occur by simply installing processes to prevent the mistakes from happening in the first place. These mistakes that happen are due to human nature and can normally not be eliminated by simple training or standard operating procedures. These steps to eliminate the defect will prevent the next step in the process from occurring if a defect is found. Normally there is some type of alert that will show there is a mistake and will fail the process from going forward. An example of a Poka Yoke would be a simple checkweigher that would kick off a package of food if it were not the correct weight.

Poka Yokes often also encompass a concept called zero quality control (ZQC). This does not mean a reduction in defects, but instead complete elimination of defects, also known as zero defects. ZQC was another concept led by the Japanese that leads to low-inventory production. The reason the inventory is so low is due to not needing excess inventory due to having to replace less defective parts as often. ZQC also focuses on quality control and data versus blaming humans on mistakes. The methodology was developed by Shigeo Shingo who knew it was human nature to make common mistakes and did not feel people should be reprimanded for them. Shingo said, "Punishment makes people feel bad, it does not eliminate defects."

This concept is important because it focuses on the customers and realizes that defects are costly, and therefore eliminating defects saves money. Many companies "rework" product to save money but do not realize to eliminate the problem in the first place. This process will eliminate rework by eliminating any defects from happening in the first place.

The first cycle of the ZQC system is the Plan–Do–Check cycle also known as PDC.

This is a traditional cycle where processes and conditions are planned out, the planned actions are performed in the Do phase, and finally quality control checks are performed in the Check phase. This method catches mistakes and also provides feedback during the Check phase. The checks in this place also account for 100% inspection; therefore, all parts or processes are looked upon indicating no defects.

There are three main types of checks or inspections that are popular:

- Judgment inspections
- Informative inspections
- Source inspections

Judgment inspections are those that are done normally by humans based on what their expectations are. They find the defect after the defect has already occurred. Informative inspections are based on statistical quality control, checks on each product, and self-checks. These inspections help reduce defects but do not eliminate them completely. Finally, the source inspections are the inspections that reduce the defects completely. Source inspections discover the mistakes before processing and then provide feedback and corrective actions so that the process has zero defects. The source inspections require 100% inspection. The feedback loop is also very quick so that there is minimal waiting time.

How to Use Poka Yokes:

Poka Yokes use two approaches:

- Control systems
- Warning systems

Control systems stop the equipment when a defect or unexpected event occurs. This prevents the next step in the process to occur so that the complete process is not performed. Warning systems signal operators to stop the process or address the issue at the time. Obviously the first of the two prevents all defects and has a more ZQC methodology because an operator could be distracted or not have time to address the problem. Control systems often also use lights or sounds to bring attention to the problem; that way the feedback loop again is very minimal.

The conclusion of Poka Yokes is to use the methodology as mistake proofing for ZQC to eliminate all defects, not just some. The types of Poka Yokes do not have to be complex or expensive, just well thought out to prevent human mistakes or accidents.

The Poka Yoke discussion stems into the correct location discussion. This technique places design and production operations in the correct order to satisfy customer demand. The concept is to increase throughput of machines ensuring the production is performed at the proper time and place. Centralization of areas helps final assemblers, but the most common practice to be effective is to unearth an effective flow. U-shaped flows normally prevent bottlenecks. Value-stream mapping is a key component in order to ensure that all steps occurring are adding value to the overall process. A reminder for value-added activities: any activity that the customer is willing to pay for. Another note to remember is to not only have a smart and efficient technique, but also only produce goods that the customer is demanding to eliminate excess inventory.

This technique is called the pull technique. Pull is the practice of not producing any goods upstream if the downstream customer does not need it. The reason this is a difficult technique is because once an efficient method is found to produce a good, the mass production begins. The operations forget if the goods are actually needed or not and begin thinking only of throughput. Even though co-manufacturing seem like a bad idea for many employers, they sometimes come in handy when a small amount of a versatile product is needed.

Push systems, on the other hand, are not effective due to predictions of customer demands.

Lean systems show the pull system utilizing machinery for 90% of requirements and limits downtime to 10% for changeovers and maintenance. This does not mean preventative maintenance should not be performed, but only that the maintenance time is reduced to 10%. Kanbans are a key factor during this Lean system in order to use a visual indicator that another part or process is required. This also prevents excess parts from being made or excess processes being performed.

Heijunka is the leveling of production and scheduling based on volume and product mix. Instead of building products according to the flow of customer orders, this technique levels the total volume of orders over a specific time so that uniform batches of different product mixes are made daily. The result is a predictable matrix of product types and volumes. For heijunka to succeed, changeovers must be managed easily. Changeovers must be as minimally invasive as possible to prevent time wasted because of product mix. Another key to heijunka is making sure that products are needed by customers. A product should not be included in a mix simply to produce inventory if it is not demanded by customers. Long changeovers should be investigated to determine the reason and devise a method to shorten them.

CONCLUSIONS

The pursuit of Lean concepts in sustainability can be effective for success. The recommended approach (Badiru and Kovach, 2012) is to implement Lean in a step-by-step approach as summarized in the following five steps:

1. Getting started – plan out the appropriate steps. This will take one to six months.

2. Create the new organization and restructure. This will take 6 to 24 months.
3. Implement Lean techniques and systems and continually improve. This will take two to four years.
4. Complete the transformation. This will take up to five years.
5. Do the entire process again to have another continuous improvement project and sustain the results.

REFERENCE

Badiru, A. B. and Kovach, Tina (2012), *Statistical Techniques for Project Control*, Taylor & Francis CRC Press, Boca Raton, FL.

Six Sigma for Sustainability

7

"When curiosity is established, the urge to learn develops."

— *Adedeji Badiru*

INTRODUCTION

Put simply, Six Sigma is about achieving uniformity in actions and products. Although it is normally applied in production settings, it has been found to be equally applicable and effective in other general applications. Sustainability activities, such as recycling, reusing, recovering, or redesigning, can benefit from the business techniques of Six Sigma. Since this is a concise FOCUS book, the full details of the techniques of Six Sigma are not provided. This chapter is designed to present a brief overview, for which further exposition can be explored in more comprehensive references. For details and examples of many of the techniques in this chapter, readers may refer to Agustiady and Badiru (2013).

Six Sigma is best defined as a business process improvement approach that seeks to find and eliminate causes of defects and errors, reduce cycle times, reduce costs of operations, improve productivity, meet customer expectations, achieve higher asset utilization, and improve return on investment. Six Sigma deals with producing data-driven results through management support of the initiatives (Agustiady and Badiru 2013). Six Sigma pertains to sustainability because without the actual data, decisions would be made on trial and error. Sustainable environments require having actual data to back up decisions so that methods are used to have improvements for future generations. The basic methodology of Six Sigma includes a five-step method a pproach that consists of the following.

Define: Initiate the project, describe the specific problem, identify the project's goals and scope, and define key customers and their Critical to Quality (CTQ) attributes.

Measure: Understand the data and processes with a view to specifications needed for meeting customer requirements, develop and evaluate measurement systems, and measure current process performance.

Analyze: Identify potential cause of problems; analyze current processes; identify relationships between inputs, processes, and outputs; and carry out data analysis.

Improve: Generate solutions based on root causes and data-driven analysis while implementing effective measures.

Control: Finalize control systems and verify long-term capabilities for sustainable and long-term success.

The goal of Six Sigma is to strive for perfection by reducing variation and meeting customer demands. The customer is known to make specifications for processes. Statistically speaking, Six Sigma is a process that produces 3.4 defects per million opportunities. A defect is defined as any event that is outside of the customer's specifications. The opportunities are considered any of the total number of chances for a defect to occur.

The Greek letter σ (sigma) marks the distance on the horizontal axis between the mean μ and the curve inflection point. The greater the distance, the greater is the spread of values encountered. The genesis of Six Sigma is represented by the curve in Figure 7.1. The figure shows a mean of 0 and a standard deviation of 1, that is, $\mu = 0$ and $\sigma = 1$. The plot also illustrates the areas under the normal curve within different ranges around the mean. The upper and lower specification limits (USL and LSL) are ±3 σ from the mean

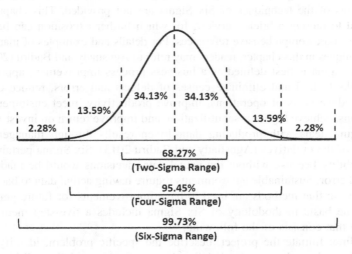

FIGURE 7.1 Areas under the Normal Curve for Six Sigma

or within a Six-Sigma spread. Because of the properties of the normal distribution, values lying as far away as ±6 σ from the mean are rare because most data points (99.73%) are within ±3 σ from the mean except for processes that are seriously out of control.

Six Sigma allows no more than 3.4 defects per million parts manufactured or 3.4 errors per million activities in a service operation. To appreciate the effect of Six Sigma, consider a process that is 99% perfect (10,000 defects per million parts). Six Sigma requires the process to be 99.99966% perfect to produce only 3.4 defects per million, that is 3.4/1,000,000 = 0.0000034 = 0.00034%. That means that the area under the normal curve within ±6 σ is 99.99966% with a defect area of 0.00034%.

The following tools are the most common Six Sigma tools and will be explained how they are to be used in the concept of sustainability.

- Project charter
- Suppliers, Inputs, Process, Outputs, and Customers (SIPOC)
- Kano model
- CTQ
- Affinity diagram
- Measurement systems analysis (MSA)
- Gage repeatability and reproducibility (Gage R&R)
- Variation
- Graphical analysis
- Location and spread
- Process capabilities
- Cause and effect (C&E) diagram
- Failure modes and effects analysis (FMEA)
- Process mapping
- Hypothesis testing
- Analysis of variance (ANOVA)
- Correlation
- Linear regression
- Theory of constraints
- Single-minute exchange of dies (SMED)
- Total productive maintenance (TPM)
- Design for Six Sigma (DFSS)
- Quality function deployment (QFD)
- DOE – design of experiments
- Control charts
- Control plan

PROJECT CHARTER

A project charter is a definition of the project that includes the following:

- Provides problem statement
- Overviews scope, participants, goals, and requirements
- Provides authorization of a new project
- Identifies roles and responsibilities

Once the project charter is approved, it should not be changed.

A project charter begins with the project name, the department of focus, the focus area, and the product or process.

A project charter serves as the focus point throughout the project to ensure the project is on track and the proper people are participating and being held accountable.

The importance of a project charter in aspect to sustainability is the living document to educate and give governance for a new project. Sustainability needs to utilize a great deal of education while giving goals and objectives. A project charter will serve as this living document for organizations with specified approaches.

SUPPLIERS, INPUTS, PROCESS, OUTPUTS, AND CUSTOMERS

The SIPOC identifies

1. Major tasks and activities
2. The boundaries of the process
3. The process outputs
4. Who receives the outputs (the customers)
5. What the customer requires of the outputs
6. The process inputs
7. Who supplies the inputs (suppliers)
8. What the process requires of the inputs
9. The best metrics to measure

Supplier – Know and work with your supplier while making your supplier improve

Input – Strive to continually improve the inputs by trying to do the right thing the first time

Process – Describe the process at a high level, but with enough detail to demonstrate to an executive or manager. Understand the process fully by knowing it 100%. Eliminate any mistakes by doing a Poka Yoke

Output – Strive to continually improve the outputs by utilizing metrics

Customer – Keep the customer's requirements in sight by remembering they are the most important aspect of the project. The customer makes the specifications, keep the CTQs of the customer in mind

SIPOC steps

1. Gain top-level view of the process
2. Identify the process in simple terms
3. Identify external inputs, such as raw materials and employees
4. Identify the customer requirements, also known as outputs
5. Make sure to include all value-added and nonvalue-added steps
6. Include both process and product output variables

The SIPOC implies that the process is understood and helps easily identify opportunities for improvement.

A SIPOC is important in concepts of sustainability because it helps develops a solution for development. Normally the process is first mapped out in a well-defined, but at a high-level.

The following is an example of a SIPOC for the steps to recycling a product:

Recycling Steps

1. Collect and process goods into a container that is friendly to deposit/refund or pick up programs
2. Undergo a recycling loop where products are sorted by type and parts
3. Parts go to a smelting plant where parts are put back to original state
4. Melted parts are sent back to a factory to make new products
5. Pieces are put back together in a completed product form at a factory
6. New products are sold
7. Supplier provides new product

The SIPOC map is continued with the outputs:

Product ready for recycling
Sorted product
Melted product
Recycled product ready for production
New product

The same process occurs with the inputs:

Newspapers
Plastic
Cardboard
Glass
Magazines
Cans
Metal

The suppliers are then identified:

ABC Local newspaper
Coca Cola bottles
Box for ABC Macaroni container
Glass jar for ABC sauce
ABC magazine
ABC canned soup

Finally, the customer is realized. It is important to understand that the customer is not always the end customer and is also part of the recycling process.

Trash recycling pickup
Container for recycling
Recycling factory
Sorting personnel
Melting personnel
Completed smelting process organizer
Delivery of recycled product to factory
Factory of reproduction
Customer buying product

The important part of a SIPOC is to look at the details of the current state and see what improvements can be made for future states. Adding specifications for any of the inputs can identify gaps in the process. Benchmarking one process to another will also identify gaps.

KANO MODEL

The Kano model was developed by Noriaki Kano in the 1980s. The Kano model is a graphical tool that further categorizes voice of the customer (VOC) and CTQs into three distinct groups:

- Must-haves
- Performance
- Delighters

The Kano model helps identify CTQs that add incremental value versus those that are simply requirements where having more is not necessarily better.

The Kano model engages customers by understanding the product attributes, which are most likely important to customers. The purpose of the tool is to support product specifications, which are made by the customer, and promote discussion while engaging team members. The model differentiates features of products rather than customer needs by understanding necessities and items that are not required whatsoever. Kano also produced a methodology for mapping consumer responses with questionnaires that focused on attractive qualities through reverse qualities. The five categories for customer preferences are as follows:

- Attractive
- One-dimensional
- Must-be
- Indifferent
- Reverse

Attractive qualities are those that provide satisfaction when fulfilled; however, do not result in dissatisfaction if not fulfilled.

One-dimensional qualities are those that provide satisfaction when fulfilled and dissatisfaction if not fulfilled.

Must-be qualities are those that are taken for granted if fulfilled but provide dissatisfaction when not fulfilled.

Indifferent qualities are those that are neither good nor bad resulting in neither customer satisfaction nor dissatisfaction.

Reverse qualities are those that result in high levels of dissatisfaction from some customers and show that most customers are not alike.

The Kano model is important to use when being sustainable because it is important to differentiate which aspects we must accomplish to protect our environment and which aspects we can gradually improve upon.

CRITICAL TO QUALITY

CTQ represents the key measurable characteristics of a product or process whose performance standards or specification limits must be met in order to satisfy the customer. These outputs represent the product or service characteristics defined by the customer (internal or external). CTQs are important to the customer. They come from the VOC. CTQs are measurable and quantifiable metrics that come from the VOC.

An affinity diagram is an organizational tool for articulating the VOCs.

Elements that are CTQ are critical to sustainability because we need to understand the critical aspects to the environment that matter most. The following are the main sources of resource consumption by rank:

- Electricity – #1 Source of resource consumption
- Natural gas – #2 Source of resource consumption
- Water/sewer – #3 Source of resource consumption

Therefore, the CTQ characteristics are electricity, natural gas, and water/sewer.

Utilizing a VOC for manufacturing internally is a good way to understand processes that the employees know a great deal about. Therefore, the production worker(s) are the customers and the questions are given to them. Based on the two questions asked and the responses, the machinery is the CTQ attribute since it affects more than one of the problems.

AFFINITY DIAGRAM

An affinity diagram is a tool conducted to place large amounts of information into an organized manner by grouping the data into characteristics. The steps for an affinity diagram are as follows:

- Step 1: Clearly define the question or focus of the exercise
- Step 2: Record all participant responses on note cards or post-it notes
- Step 3: Lay out all note cards or post the post-it notes onto a wall
- Step 4: Look for and identify general themes
- Step 5: Begin moving the note cards or post-it notes into the themes until all responses are allocated
- Step 6: Reevaluate and make adjustments

The exact same methodology for a basic process can be done for a manufacturing or business process where sustainability is in question. The pros and cons are then sought after to have a decision. The decision should be made by having a consensus from the group where the pros outweigh the cons.

MEASUREMENT SYSTEMS ANALYSIS

Gage R&R

Gage R&R is a MSA technique that uses continuous data based on the following principles:

- Data must be in statistical control
- Variability must be small compared to product specifications
- Discrimination should be about one-tenth of the product specifications or process variations
- Possible sources of process variation are revealed by measurement systems
- Repeatability and reproducibility are primary contributors to measurement errors
- The total variation is equal to the real product variation plus the variation due to the measurement system
- The measurement system variation is equal to the variation due to repeatability plus the variation due to reproducibility
- Total (observed) variability is an additive of product (actual) variability and measurement variability

Discrimination is the number of decimal places that can be measured by the system. Increments of measure should be about one-tenth of the width of a product specification or process variation that provides distinct categories.

Accuracy is the average quality near to the true value.

The *true value* is the theoretically correct value.

Bias is the distance between the average value of the measurement and the true value, the amount by which the measurement instrument is consistently off target, or systematic error. *Instrument accuracy* is the difference between the observed average value of measurements and the true value. Bias can be measured based on instruments or operators. Operator bias occurs when different operators calculate different detectable averages for the same measure. Instrument bias results when different instruments calculate different detectable averages for the same measure.

Precision encompasses total variation in a measurement system, the measure of natural variation of repeated measurements, and repeatability and reproducibility.

Repeatability is the inherent variability of a measurement device. It occurs when repeated measurements are made of the same variable under absolutely identical condition (same operators, setups, test units, environmental conditions) in the short term. Repeatability is estimated by the pooled standard deviation of the distribution of repeated measurements and is always less than the total variation of the system.

Reproducibility is the variation that results when measurements are made under different conditions. The different conditions may be operators, setups, test units, or environmental conditions in the long term. Reproducibility is estimated by the standard deviation of the average of measurements from different measurement conditions.

The *measurement capability index* is also known as the precision-to-tolerance (P/T) ratio. The equation is P/T = 5.15 × σMS)/tolerance. The P/T ratio is usually expressed as a percentage and indicates what percentage of tolerance is taken up by the measurement error. It considers both repeatability and the reproducibility. The ideal ratio is 8% or less; an acceptable ratio is 30% or less. The 5.15 standard deviation accounts for 99% of Mean Squares (MS) variation and is an industry standard.

The P/T ratio is the most common estimate of measurement system precision. It is useful for determining how well a measurement system can perform with respect to the specifications. The specifications, however, may be inaccurate or need adjustment. The %R&R = (σMS/σTotal) × 100 formula addresses the percentage of the total variation taken up by measurement error and includes both repeatability and reproducibility.

A Gage R&R can also be performed for discrete data, which is also known as binary data. These data are also known as yes/no or defective-/nondefective-type data. The data still require at least 30 data points. The percentages of repeatability, reproducibility, and compliance should be measured. If no repeatability can be shown, there will also be no reproducibility. The matches should be above 90% for the evaluations. A good measurement system will have a 100% match for repeatability, reproducibility, and compliance.

If the result is below 90%, the operational definition must be revisited and redefined. Coaching, teaching, mentoring, and standard operating procedures should be reviewed, and the noise should be eliminated. A decision needs to be made on which activities and production equipment are sustainable.

The Gage R&R bars are desired to be as small as possible, driving the part-to-part bars to be larger.

The averages of each operator is different, meaning the reproducibility is suspect. The operator is having a problem making consistent measurements.

The Operator*Samples interactions lines should be reasonably parallel to each other. The operators are not consistent to each other.

The measurement by samples graph shows that there is minimal spread for each sample and a small amount of shifting between samples.

The sample times operator of 0.706 show that the interaction was not significant, which is what is wanted from this study.

The percentage contribution part to part of 10.81 shows the parts are the same.

The total Gage R&R % study variation of 94.44, percentage contribution of 89.19, tolerance of 143.25, and distinct categories of 1 showed this there was no repeatability, reproducibility, and was not a good Gage. The number of categories being less than 2 shows the measurement system is of minimal value since it will be difficult to distinguish one part from another.

The gage run chart shows that there is no consistency between measurements.

Conclusion, B1 is the best blender, and Dominic is the best operator. There is no reproducibility or repeatability between any of the measurements.

VARIATION

Variation is present in all processes, but the goal is to reduce the variation while understanding the root cause of where the variation comes from in the process and why. For Six Sigma to be successful, the processes must be in control statistically and the processes must be improved by reducing the variation. The distribution of the measurements should be analyzed to find the variation and depict the outliers or patterns.

The study of variation began with Dr. W. E. Deming, who was also known as the Father of Statistics. Deming stated that variation happens naturally, but the purpose is to utilize statistics to show patterns and types of variations. There are two types of variations that are sought after: special cause variation and common cause variation. Special cause variation refers to out-of-the-ordinary events, such as a power outage. Whereas common cause variation is inherent in all processes and is typical. The variation is sought to be reduced so that the processes are predictable, in statistical control, and have a known process capability. A root cause analysis should be performed on special cause variation so that the occurrence does not happen again. Management is in charge of common cause variation where action plans are given to reduce the variation.

Assessing the location and spread are important factors as well. Location is known as the process being centered along with the process requirements. Spread is known as the observed values compared to the specifications. The stability of the process is required. The process is said to be in statistical control if the distribution of the measurements have the same shape, location, and spread over time. This is the point in time where all special causes of variation are removed and only common cause variation is present.

An *average, central tendency* of a data set is a measure of the "middle" or "expected" value of the data set. Many different descriptive statistics can be chosen as measurements of the central tendency of the data items. These include the arithmetic mean, the median, and the mode. Other statistical measures such as the standard deviation and the range are called measures of spread of data. An average is a single value meant to represent a list of values. The most common measure is the arithmetic mean, but there are many other measures of central tendency such as the median (used most often when the distribution of the values is skewed by small numbers with very high values).

As stated before, special cause variation would be occurrences such as power outages and large mechanical breakdowns. Common cause variations would be occurrences such as electricity being different by a few thousand kilowatts per month. In order to understand the variation, graphical analyses should be done followed by capability analyses.

It is important to understand the variation in the systems so that the best performing equipment is used. The variation sought after is in turn utilized for sustainability studies. The best performing equipment should be utilized the most and the least performing equipment should be brought back to its original state of condition and then upgraded or fixed to be capable. Capability indices are explained next.

PROCESS CAPABILITIES

The capability of a process is the spread that contains most of the values of the process distribution. Capability can only be established on a process that is stable with a distribution that only has common cause variation.

Capable Process (C_p)

A process is capable ($C_p \geq 1$) if its natural tolerance lies within the engineering tolerance or specifications. The measure of process capability of a stable process is $6\hat{\sigma}$, where $\hat{\sigma}$ is the inherent process variability that is estimated from the process. A minimum value of $C_p = 1.33$ is generally used for an ongoing process. This ensures a very low reject rate of 0.007% and therefore is an effective strategy for prevention of nonconforming items. C_p is defined mathematically as

$$C_p = \frac{\text{USL} - \text{LSL}}{6\sigma} = \frac{\text{allowable process spread}}{\text{actual process spread}},$$

where:

USL = upper specification limit and
LSL = lower specification limit.

C_p measures the effect of the inherent variability only. The analyst should use R-bar/d_2 to estimate $\hat{\sigma}$ from an R-chart that is in a state of statistical control, where R-bar is the average of the subgroup ranges and d_2 is a normalizing factor that is tabulated for different subgroup sizes (n). We do not have to verify control before performing a capability study. We can perform the study, then verify control after the study with the use of control charts. If the process is in control during the study, then our estimates of capabilities are correct and valid. However, if the process was not in control, we would have gained useful information, as well as proper insights as to the corrective actions to pursue.

Capability Index (C_{pk})

Process centering can be assessed when a two-sided specification is available. If the capability index (C_{pk}) is equal to or greater than 1.33, then the process may be adequately centered. C_{pk} can also be employed when there is only one-sided specification. For a two-sided specification, it can be mathematically defined as

$$C_{pk} = \text{Minimum} \left\{ \frac{\text{USL} - \bar{X}}{3\hat{\sigma}}, \quad \frac{\bar{X} - \text{LSL}}{3\hat{\sigma}} \right\},$$

where

$\bar{X}$ = overall process average.

However, for a one-sided specification, the actual C_{pk} obtained is reported. This can be used to determine the percentage of observations out of specification. The overall long-term objective is to make C_p and C_{pk} as large as possible by continuously improving or reducing process variability, $\hat{\sigma}$, for every iteration so that a greater percentage of the product is near the key quality characteristics target value. The ideal is to center the process with zero variability.

If a process is centered but not capable, one or several courses of action may be necessary. One of the actions may be that of integrating designed experiment to gain additional knowledge on the process and in designing control strategies. If excessive variability is demonstrated, one may conduct a nested design with the objective of estimating the various sources of variability. These sources of variability can then be evaluated to determine what strategies to use in order to reduce or permanently eliminate them. Another action may be that of changing the specifications or continuing production and then sorting the items. Three characteristics of a process can be observed with respect to capability, as summarized next.

1. The process may be centered and capable.
2. The process may be capable but not centered.
3. The process may be centered but not capable.

GRAPHICAL ANALYSIS

Graphical analyses are visual representations of tools that show meaningful key aspects of projects. These tools are commonly known as dotplots, histograms, normality plots, Pareto diagrams, second-level Pareto diagrams (also known as stratification), boxplots, scatter plots, and marginal plots. The plotting of data is a key beginning step to any type of data analysis because it is a visual representation of the data.

If a particular manufacturing company wants to understand where majority of their electrical costs are coming from while trying to reduce those costs, the following steps are to be followed.

PROCESS MAPPING

The importance of process mapping is to depict all functions in the process flow while understanding if the functions are value added or non-value added. Any delays are to be eliminated and decisions are meant to be as efficient as possible. The purpose of process mapping is to have a visual image of the process.

C&E DIAGRAM

After a process is mapped, the C&E diagram can be completed. This process is so important because it completes the root cause analysis. The basis behind root cause analysis is to ask, "Why?" five times in order to get to the actual root cause. Many times problems are "bandaided" in order to fix the top-level problem, but the actual problem itself is not addressed.

The fishbone is broken out to the most important categories in an environment:

- Measurements
- Material
- Personnel
- Environment
- Methods
- Machines

This process requires a team to do a great deal of brainstorming where they focus on the causes of the problems based on the categories. The "fish head" is the problem statement.

FAILURE MODE AND EFFECT ANALYSIS

In order to select action items from the C&E diagram and prioritize the projects, FMEAs are completed. The FMEA will identify the causes, assess risks, and determine further steps. The steps to an FMEA are the following:

1. Define process steps.
2. Define functions.
3. Define potential failure modes.
4. Define potential effects of failure.
5. Define the severity of a failure.
6. Define the potential mechanisms of failure.
7. Define current process controls.
8. Define the occurrence of failure.
9. Define current process control detection mechanisms.
10. Define the ease of detecting a failure.

11. Multiply severity, occurrence, and detection to calculate a risk priority number (RPN).
12. Define recommended actions.
13. Assign actions with key target dates to responsible personnel.
14. Revisit the process after actions have been taken to improve it.
15. Recalculate RPNs with the improvements.

What can be seen from the FMEA that is an important aspect to sustainability is the RPN number reducing after the action items. It is important to understand the process' severity to a customer and increasing the capability of the process to in turn improve the process. The RPNs reducing will make the entire process more sustainable by being able to deliver the process at the best capabilities through thorough project management. It is important to maintain the FMEA so that once a process is improved it is not forgotten about.

HYPOTHESIS TESTING

Hypothesis testing validates assumptions made by verification of the processes based on statistical measures. It is important to use at least 30 data points for hypothesis testing so that there are enough data to validate the results.

Normality of the data points must be found in order for the hypothesis testing to be accurate.

The assumptions are shown in the null and alternate hypothesis:

H_0 = (the null hypothesis): The difference is equal to the chosen reference value $\mu1 - \mu2 = 0$.

H_a = (the alternate hypothesis): The difference is not equal to the chosen reference value $\mu1$-$\mu2$ is not = 0.

95% confidence interval (CI) for mean difference: (1.16, 6.69) t-test of mean difference = 0 (versus not = 0): t-value = 2.90; p-value = 0.007.

The CI for the mean difference between the two materials does not include zero, which suggests a difference between them. The small p-value ($p = 0.007$) further suggests that the data are inconsistent with H_0: $\mu\ d = 0$, that is, the two materials do not perform equally. Specifically, the first set (mean = 79.697) performed better than the next set (mean = 83.623) in terms of weight control over the time span. Conclusion, Reject H_0, the difference is not equal to the chosen reference value: $\mu1 - \mu2$ is not = 0. The Histogram of Differences and the Boxplot of the confidence interval for the mean difference between the two materials does not include zero, which suggests a difference between them. The small p-value (p = 0.007) further suggests that the data are inconsistent with *H0*: μ d = 0, that is, the two materials do not perform equally. Specifically, the first set (mean = 79.697) performed better than the other set (mean = 83.623) in terms of weight control over the time span. Conclusion, Reject *Ho*, the difference is not equal to the chosen reference value: $\mu1 - \mu2$ is not =0.

ANALYSIS OF VARIANCE

The purpose of an ANOVA, also known as analysis of variance, is to determine if there is a relationship between a discrete, independent variable and a continuous, dependent output. There is a one-way ANOVA, which includes one-factorial variance, and a two-way ANOVA, which includes a two-factorial variance. Three sources of variability are sought after:

Total – Total variability within all observations

Between – Variation between subgroup means

Within – Random chance variation within each subgroup, also known as noise

The equation for a one-way ANOVA is as follows:

$$SS_T = SS_F + SS_e.$$

The principles for the one-way ANOVA and two-way ANOVA are the same, except that in a two-way ANOVA the factors can take on many levels. The total variability equation for a two-way ANOVA is as follows:

$$SS_T = SS_A + SS_B + SS_{AB} + SS_e,$$

where

SS_T = total sum of squares,

SS_F = sum of squares of the factor,

SS_e = sum of squares from error,

SS_A =sum of squares for factor A

SS_B = sum of squares for factor *B*, and

SS_{AB} = sum of squares due to interaction of factors *A* and *B*,

If the ANOVA shows that at least one of the means is different, a pairwise comparison is done to show which means are different. The residuals, variance, and normality should be examined and the main effects plot and interaction plots should be generated.

The F-ratio in an ANOVA compares the denominator to the numerator to see the amount of variation expected. When the F-ratio is small, which is normally close to 1, the value of the numerator is close to the value of the denominator, and the null hypothesis cannot be rejected stating the numerator and denominator are the same. A large F-ratio indicates the numerator and denominator are different, also known as the MS Error, where the null hypothesis is rejected.

Outliers should also be sought after in the ANOVA showing the variability is affected.

The main effects plot shows the mean values for the individual factors being compared. The differences between the factor levels can be seen with the slopes in the lines. The p-values can help determine if the differences are significant.

Interaction plots show the mean for different combinations of factors.

CORRELATION

The linear relationship between two continuous variables can be measured through correlation coefficients. The correlation coefficients are values between −1 and 1.

If the value is around 0, there is no linear relationship.
If the value is less than 0.05, there is a weak correlation.
If the value is less than 0.08, there is a moderate correlation.
If the value is greater than 0.08, there is a strong correlation.
If the value is around 1, there is a perfect correlation.

SIMPLE LINEAR REGRESSION

The regression analysis describes the relationship between a dependent variable and independent variable as a function $y = f(x)$.

The equation for simple linear regression as a model is as follows:

$$Y = b_0 + b_1 x + {}_E ,$$

where
   Y is the dependent variable;
   b_0 is the axis intercept;
   b_1 is the gradient of the regression line;
   x is the independent variable; and

$_E$ is the error term or residuals.

The predicted regression function is tested with the following formula:

$$R^2 = \frac{SSTO - SSE}{SSTO},$$

where

$$SSTO = \begin{cases} Y' - n\bar{Y}^2 & \text{if constant} \\ \underline{Y}\underline{Y} & \text{if no constant} \end{cases}.$$

Note: When the no constant option is selected, the total sum of square is uncorrected for the mean. Thus, the R^2 value is of little use, since the sum of the residuals is not zero.

The F-test shows if the predicted model is valid for the population and not just the sample. The model is statistically significant if the predicted model is valid for the population.

The regression coefficients are tested for significance through t-tests with the following hypothesis:

H_o: $b_0 = 0$, the line intersects the origin.

H_A: $b_0 \neq 0$, the line does not intersect the origin.

H_0: $b_1 = 0$, there is no relationship between the independent variable xi, and the dependent variable y.

H_A: $b_1 \neq 0$, there is a relationship between the independent variable xi, and the dependent variable y.

After the inverse relationship is seen, a regression analysis can be performed.

An example is shown next for the analysis of whether there was a pressure degradation over time on a particular piece of equipment.

A linear relationship was sought after. First, it was sought to see if there was a correlation since it can be seen that there is a linear relationship between the variables. The y was the measurement and the x was the time.

THEORY OF CONSTRAINTS

Dr. Eliyahu M. Goldratt created a theory of constraints (TOC). This management theory proved that every system has at least one constraint, limiting it from 100% efficiency. The analysis of a system will show the boundaries of the system. TOC not only shows the cause of the constraints, but it also provides a way to resolve the constraints. There are two underlying concepts with TOC:

1. System as chains
2. Throughput, inventory management, and operating expenses

The performance of the entire system is called the chain. The performance of the system is based on the weakest link of the chain or the constraint. The remaining links are known as nonconstraints. Once the constraint is improved, the system becomes more productive or efficient, but there is always a weakest link or constraint. This process continues until there is 100% efficiency.

If there are three manufacturing lines and they produce the following:

1. 250 units/day
2. 500 units/day
3. 600 units/day

The weakest link is manufacturing line 1 because it produces the least amount of units/day. The weakest link is investigated until it reaches the capacity of the nonconstraints. After the improvement has been made, the new weakest link is investigated until the full potential of the manufacturing lines can be fulfilled without exceeding market demand. If the external demand is fewer than the internal capacity, it is known as an external constraint.

THROUGHPUT

Throughput can be defined as (sales price – variable cost)/time. Profits should be understood when dealing with throughput. Inventories are known as raw materials, unfinished goods, purchased parts, or any investments made. Inventory should be seen as dollars on shelves. Any inventory is a waste, unless utilized in a just in time manner.

Operating expenses should include all expenses utilized to produce a good. The less the operating expenses, the better. These costs should include direct labor, utilities, supplies, and depreciation of assets.

Applying the TOC concept helps guide making the weakest link stronger. There are five steps to the process of TOC:

1. Identify the constraint or the weakest link.
2. Exploit the constraint by making it as efficient as possible without spending money on the constraint or considering upgrades.
3. Subordinate everything else to the constraint – adjust the rest of the system so the constraint operates at its maximum productivity.

Evaluate the improvements to ensure the constraint has been addressed properly and it is no longer the constraint. If it is still the constraint, complete the steps, otherwise skip step 4.

4. Elevate the constraint – this step is only required if steps two and three were not successful. The organization should take any action on the constraint to eliminate the problem. This is the process where money should be spent on the constraint or upgrades should be investigated.

5. Identify the next constraint and begin the five-step process over. The constraint should be monitored and continuous improvement should be completed.

SINGLE-MINUTE EXCHANGE OF DIES

What Is SMED?

SMED consists of the following:

- Theory and set of techniques to make it possible to perform equipment setup and changeover operations in under 10 minutes
- Originally developed to improve die press and machine tool setups, but principles apply to changeovers in all processes
- It may not be possible to reach the "single-minute" range for all setups, but SMED dramatically reduces setup times in almost EVERY case
- Leads to benefits for the company by giving customers variety of products in just the quantities they need
- High quality, good price, speedy delivery, less waste, and cost effective

It is important to understand that large lot production leads to trouble.

The three key topics to consider when understanding large lot production are the following:

- Inventory waste
  - Storing what is not sold costs money
  - Ties up company resources with no value to the product
- Delay
  - Customers have to wait for the company to produce entire lots rather than just what they want

- Declining quality
  - Storing unsold inventory increases chances of product being scrapped or reworked, thus adding costs

Once this is realized, the benefits of SMED can be understood:

- Flexibility
  - Meet changing customer needs without excess inventory
- Quicker delivery
  - Small-lot production equals less lead time and less customer waiting time
- Better quality
  - Less inventory storage equals fewer storage-related defects
  - Reduction of setup errors and elimination of trial runs for new products
- Higher productivity
  - Reduction in downtime
  - Higher equipment productivity rate

Two types of operations are realized during setup operations which consist of internal and external operations. Internal setup is a setup that can only be done when the machine is shut down (i.e., a new die can only be attached to a press when the press is stopped).

External setup is a setup that can be done while the machine is still running (i.e., bolts attached to a die can be assembled and sorted while the press is operating).

It is important to convert as much internal work as possible to external work.

Four important questions to ask yourself when understanding SMED are the following:

- How might SMED benefit your factory?
- Can you see SMED benefiting you?
- What operations are internal operations?
- What operations are external operations?

There are three stages to SMED, which are defined as follows:

- Separate internal and external setup
  - Distinguish internal versus external

  – By preparing and transporting while the machine is running can cut changeover times by as much as 50%

- Convert internal setup to external setup
  – Re-examine operations to see whether any steps are wrongly assumed as internal steps
  – Find ways to convert these steps to external setups
- Streamline all aspects of setup operations
  – Analyze steps in detail
  – Use specific principles to shorten time needed especially for steps internally with machine stopped

Five Traditional Setups steps are also defined:

- **Preparation** – Ensures that all the tools are working properly and are in the right location
- **Mounting and Extraction** – Involves the removal of the tooling after the production lot is completed and the placement of the new tooling before the next production lot
- **Establishing Control Settings** – Setting all the process control settings prior to the production run. Inclusive of calibrations and measurements needed to make the machine tooling operate effectively
- **First Run Capability** – This includes the necessary adjustments (recalibrations, additional measurements) required after the first trial pieces are produced
- **Setup Improvement**– The time after processing during which the tooling machinery is cleaned, identified, and tested for functionality prior to storage

The three stages of SMED are explained next.

DESCRIPTION OF STAGE 1 – SEPARATE INTERNAL VERSUS EXTERNAL SETUP

Three techniques help us separate internal versus external setup tasks:

1. Use checklists
2. Perform function checks
3. Improve transport of die and other parts

Checklists: A checklist lists everything required to set up and run the next operation. The list includes the following items:

- Tools, specifications, and workers required
- Proper values for operating conditions such as temperature and pressure
- Correct measurement and dimensions required for each operation
- Checking item of the list before the measurement configuration m/c is stopped helps prevent mistakes that come up after internal setup begins

Function Checks

- Should be performed before setup begins so that repair can be made if something does not work right
- If broken dies, molds, or jigs are not discovered until test runs are done, a delay will occur in internal setup
- Make sure such items are in working order before they are mounted and it will cut down setup time a great deal

Improved Transport of Parts and tools

- Dies, tools, jigs, gauges, and other items needed for an operation must be moved between storage areas and machines, then back to storage once a lot is finished
- To shorten the time the machine is shut down, transport of these items should be done during external setup
- In other words, new parts and tools should be transported to the machine before the machine is shut down for changeover

DESCRIPTION OF STAGE 2 – CONVERT INTERNAL SETUPS TO EXTERNAL SETUPS

Advance Preparation of Conditions

- Get necessary parts, tools, and conditions ready before internal setup begins
- Conditions such as temperature, pressure, or position of material

can be prepared externally while the machine is running (i.e., preheating of mold/material)

Function Standardization

- It would be expensive and wasteful to make external dimensions of every die, tool, or part the same, regardless of the size or shape of the product it forms. Function standardizations avoid this waste by focusing on standardizing only those elements whose functions are essential to the setup
- Function standardization might apply to dimensioning, centering, securing, expelling, or gripping

Implementing Function Standardization with Two Steps

- Look closely at each individual function in your setup process and decide which functions, if any, can be standardized
- Look again at the functions and think about which can be made more efficient by replacing the fewest possible parts (i.e., clamping function standardization)

DESCRIPTION OF STAGE 3 – STREAMLINE ALL ASPECTS OF THE SETUP OPERATION

- External setup improvement includes streamlining the storage and transport of parts and tools
- In dealing with small tools, dies, jigs, and gauges, it is vital to address issues of tool and die management

Ask Q's Like (Ask Questions Like)

- What is the best way to organize these items?
- How can we keep these items maintained in perfect condition and ready for the next operation?
- How many of these items should we keep in stock?

Improving Storage and Transport

- Operation for storing and transporting dies can be very time-consuming, especially when your factory keeps a large number of dies on hand
- Storage and transport can be improved by marking the dies with color codes and location numbers of the shelves where they are stored

Streamlining Internal Setup

- Implement parallel operations, using functional clamps, eliminating adjustments, and mechanization

Implementing Parallel Operations

Machines such as plastic molding machines and die costing machines often require operation at both the front and back of the machine. One person changeovers of such machines mean wasted time and movement because the same person is constantly walking back and forth from one end of the machine to the other.

Parallel operations divide the setup operation between two people, one at each end of the machine. When setup is done using parallel operations, it is important to maintain reliable and safe operations and minimize waiting time. To help streamline parallel operations, workers should develop and follow procedural charts for each setup.

The final understanding of SMED comes from basic principles such as observing with videos.

If there is nobody in the screen, it means there is waste present.

It is important to understand that SMED is more than just a series of techniques. It is a fundamental approach to improvement activities. A personal action plan should be found to adhere to each business's needs. It important to find ways to implement SMED into environments to continue the sustainability of the businesses. To begin the process, a communication plan should be implemented.

TPM – TOTAL PRODUCTIVE MAINTENANCE

TPM has been a well-known activity that has several names associated within. Many people associate TPM with Total Predictive Maintenance or

Total Preventative Maintenance. The association explained here will be total productive maintenance, but includes the earlier explanations as well.

TPM is performed in the improve phase based on downtimes or efficiency losses. The downtimes associated can be planned or unplanned. The goal of TPM is to increase all operational equipment efficiencies to above 85% by eliminating any wasted time such as setup time (see SMED section), idle times, downtimes, startup delays, and any quality losses.

TPM ensures minimal downtime, but in turn requires no defects as well. There are three basic steps for TPM that have several steps within each.

1. Analyze the current processes
    a. Calculate any costs associated with maintenance
    b. Calculate Overall Equipment Effectiveness (OEE) by finding the proportion of quality products produced at a given line speed

2. Restore equipment to its original and high operating states
    a. Inspect the machinery
    b. Clean the machinery
    c. Identify necessary repairs on the machinery
    d. Document defects
    e. Create a scheduling mechanism for maintenance
    f. Ensure maintenance has repaired machinery and improvements are sustained

3. Preventative maintenance to be carried out
    a. Create a schedule for maintenance with priorities – include high machinery defects, replacement parts, and any information pertaining to them.
    b. Create stable operations – complete the root cause analysis on high machinery defects and machinery that causes major downtime.
    c. Create a planning and communication system – documentation of preventative maintenance activities should be accessible to all people so planning and prioritization within is completed.
    d. Create processes for continuous maintenance – inspections should occur regularly and servicing for any machinery should be noted on a scheduled basis.
    e. Internal operations should be optimized – any internal operations should be benchmarked with improvements from other areas to eliminate time spent on root cause analysis. When defects of machinery are not understood, it is important to put the machinery back to its original state to understand the root causes more efficiently. Time to exchange parts or retrieve parts should also be minimized.

    f. Continuous improvement on preventative maintenance – train employees for early detection of problems and maintenance measures. Visual controls should be put in place for changeovers. The 5S should take place to eliminate wasted time. The documentation should be communicated and plans should be presented regularly. All aspects should be looked upon to see if continuous improvements can be made.

The key TPM indicators will be able to show the following main issues:

- OEE
- Mean time between failures
- Mean time to repair

TPM is crucial to sustainability because it involves all the employees, including high-level managers, and creates planning for preventative maintenance so the issues are fixed before they become an error or defect. TPM is also a journey for educating and training the workforce to be familiar with the machinery, parts, processes, and damages while being productive.

DESIGN FOR SIX SIGMA

DFSS is another process that is included in a phase called DMADV, which stands for the following:

Define
Measure
Analyze
Design
Verify

The difference of DMADV from Define, Measure, Analyze, Improve, and Control (DMAIC) is the design and verification portions. DMAIC is process improvement-driven, whereas DMADV is for designing new products or services. Design stands for the designing of new processes required, including the implementation.

    Verify stands for the results being verified and the performance of the design to be maintained.

The purpose of DFSS is very similar to the regular DMAIC cycle where it is a customer-driven design of processes with Six Sigma capabilities. DFSS does not only have to be manufacturing-driven, the same methodologies can be used in service industries. The process is top down with flow down CTQs that match flow up capabilities. DFSS is quality-based where predictions are made regarding first pass quality. The quality measurements are driven through predictability in the early design phases. Process capabilities are utilized to make final design decisions.

Finally, process variances are monitored to verify Six Sigma customer requirements are met.

The main tools utilized in DFSS are FMEAs, QFD, DOE, and simulations.

QUALITY FUNCTION DEPLOYMENT

Dr. Yoji Akao developed QFD, in 1966, in Japan. There was a combination of quality assurance and quality control that led to value engineering analyses. The methods for QFD are simply to utilize consumer demands into designing quality functions and methods to achieve quality into subsystems and specific elements of processes. The basis for QFD is to take customer requirements from the VOC and relaying them into engineering terms to develop products or services. Graphs and matrices are utilized for QFD. A house-type matrix is compiled to ensure that the customer needs are being met into the transformation of the processes or services designed. The QFD house is a simple matrix where the legend is used to understand quality characteristics, customer requirements, and completion.

DESIGN OF EXPERIMENTS

DOEs is an experimental design that shows what is useful, what is a negative connotation, and what has no effect. Majority of the time, 50% of the designs have no effect.

DOEs require a collection of data measurements, systematic manipulation of variables, also known as factors, placed in a prearranged way (experimental designs), and control for all other variables. The basis behind DOEs is to test everything in a prearranged combination and measure the effects of each of the interactions.

The following DOE terms are used:

- Factor: An independent variable that may affect a response
- Block: A factor used to account for variables that the experimenter wishes to avoid or separate out during analysis
- Treatment: Factor levels to be examined during experimentation
- Levels: Given treatment or setting for an input factor
- Response: The result of a single run of an experiment at a given setting (or given combination of settings when more than one factor is involved)
- Replication (Replicate): Repeated run(s) of an experiment at a given setting (or given combination of settings when more than one factor is involved)

There are two types of DOEs: full-factorial design and fractional factorial design.

Full-factorial DOEs determine the effect of the main factors and factor interactions by testing every factorial combination.

A full-factorial DOE factors all levels combined with one another covering all interactions. The effects from the full-factorial DOE can then be calculated and sort into main effects and effects generated by interactions.

Effect = Mean Value of Response when Factor Setting is at High Level $(Y_A +)$ – Mean Value of Response when Factor Setting is at Low Level (Y_A-)

In a full-factorial experiment, all of the possible combinations of factors and levels are created and tested.

In a two-level design (where each factor has two levels) with k factors, there are 2^k possible scenarios or treatments.

- Factors each with two levels, we have $2^2 = 4$ treatments
- Factors each with two levels, we have $2^3 = 8$ treatments
- k factors each with two levels, we have 2^k treatments

The analysis behind the DOE consists of the following steps:

1. Analyze the data
2. Determine factors and interactions
3. Remove statistically insignificant effects from the model such as p-values of less than 0.1 and repeat the process
4. Analyze residuals to ensure the model is set correctly
5. Analyze the significant interactions and main effects on graphs while setting up a mathematical model

6. Translate the model into common solutions and make sustainable improvements

A fractional factorial design locates the relationship between influencing factors in a process and any resulting processes while minimizing the number of experiments. Fractional factorial DOEs reduce the number of experiments while still ensuring that the information lost is as minimal as possible. These types of DOEs are used to minimize time spent, money spent, and eliminate factors that seem unimportant.

The formula for a fractional factorial DOE is as follows:

2^{k-q}, where q equals the reduction factor.

The fractional factorial DOE requires the same number of positive and negative signs as a full-factorial DOE.

CONTROL CHARTS

Control charts are a great interpretation of understanding whether projects are being sustained by utilizing process monitoring. The process spread can be understood through control charts, while also interpreting whether the process is in control and predictable. Common cause and special cause variation can be found through control charts. The amount of samples taken is an important aspect to control charts along with the frequency of sampling. It is important to have a random yet normal pattern of data. For example, if data are taken during normal operating conditions and then one data point is taken during a changeover, the data will be skewed and show a point out of control.

X-Bar and Range Charts

The R-chart is a time plot useful for monitoring short-term process variations. The X-bar chart monitors longer-term variations where the likelihood of special causes is greater over time. Both charts utilize control lines called upper and lower control limits and central lines; both types of lines are calculated from process measurements. They are not specification limits or percentages of the specifications or other arbitrary lines based on experience. They represent what a process is capable of doing when only common cause variation exists, in which case the data will continue to fall in a random fashion within the control limits and the process is in a state of statistical control. However, if a special cause acts on a process, one or more data points will be outside the control limits and the process will no longer be in a state of statistical control.

The following components should be used for control chart purposes:

- **UCL** – Upper control limit
- **LCL** – Lower control limit
- **CL** – Center Line – shows where the characteristic average falls
- **USL** – Upper specification limit or upper customer requirement
- **LSL** – Lower specification limit or lower customer requirement

Control limits describe the stability of the process. Specifically, control limits identify the expected limits of normal, random, or chance variation that is present in the process being monitored. Control limits are set by the process.

Specification limits are those limits that describe the characteristics the product or process must have in order to conform to customer requirements or to perform properly in the next operation.

Control limits describe the representative nature of a stable process. Specifically, control limits identify the expected limits of normal, random, or chance variation that is present in the process being monitored. Sustainable processes must follow these rules.

CONTROL PLANS

A control plan is a vital part of sustainability because without it, there is no sustainability. A control plan takes the improvements made and ensures that they are being maintained and continuous improvement is achieved. A control plan is a very detailed document that includes who, what, where, when, and why (the why is based on the root cause analysis). The 12 basic steps of a control plan are as follows:

1. Collect existing documentation for the process
2. Determine the scope of the process for the current control plan
3. Form teams to update the control plan regularly
4. Replace short-term capability studies with long-term capability results
5. Complete control plan summaries
6. Identify missing or inadequate components or gaps
7. Review training, maintenance, and operational action plans
8. Assign tasks to team members

9. Verify compliance of actual procedures with documented procedures
10. Retrain operators
11. Collect sign-offs from all departments
12. Verify effectiveness with long-term capabilities

CONCLUSIONS

A control plan ensures consistency, while eliminating as much variation from the system as possible. The plans are essential to operators because it enforces standard operating procedures and eliminates changes in processes. It also ensures PMs process management activities are performed and the changes made to the processes are actually improving the problem that was found through the root cause analysis. Control plans hold people accountable if reviewed at least quarterly. As an example, a house is designed for operational stability and sustainability that encompass the topics discussed earlier. The house can be rearranged or reworked with different goals and tools set out for the particular business or manufacturing environment. The base of the house shows the stability and the areas needed for a stable and sustainable work environment. The pillars are the goals and tools utilized within the house in order for the roof to be stable upon the pillars. The roof of the house is the ultimate goal for a beneficial environment. Sustainability provides the foundation for achieving structural longevity in our global infrastructure.

REFERENCE

Agustiady, Tina and Badiru, Adedeji B. (2013), *Sustainability: Utilizing Lean Six Sigma Techniques*, Taylor & Francis CRC Press, Boca Raton, FL.

Workforce Development for Sustainability during COVID-19

8

"Develop, retain, preserve, and sustain the workforce for the knowledge economy."

—*Adedeji Badiru*

INTRODUCTION

This chapter is based on Badiru and Barlow (2020a, 2020b). Charity begins at home, so the saying goes. Conversely, the reversibility of practices can begin at work. The issue of workforce development (WFD) is typically seen as a work pursuit. The operational fact is that what we learn at work can inform what we practice at home and elsewhere. Thus, the presentation in this chapter provides a linkage between recommended WFD strategies and desired sustainability practices.

Due to the human resource ravages of COVID-19, the workforce in every part of the world is facing unprecedented challenges. New approaches to human resource management (HRM) and WFD are urgently needed.

A well-developed and versatile workforce is the lifeline for not only the operational buoyancy of commercial organizations, business, and industry but also the economic vitality of government organizations, military systems, political states, regions, and nations. Inasmuch as every organization thrives on human resources, it is essential that we take a new contemporary look at the processes of WFD within and after the ongoing COVID-19 pandemic. This chapter uses a mix of analytical systems engineering concepts and social science theory to present a composite methodology for WFD in an era of COVID-19. It is expected that this new methodology could have an immediate impact on business, industry, academia, and even on government. Global economic development is predicated on WFD. Consequently, we are all in this challenge together.

The world today is more interconnected than ever before. The fast pace and vast reach of travel, coupled with effortless Internet connectivity, have created systematic linkages throughout the world. As we have seen in recent events of COVID-19 pandemic and social protests, issues in one part of the world can quickly and easily instigate issues in other parts of the world. Thus, it is more of a systems world now than ever before. As a result of the global lockdown response to COVID-19, what we would have seen as an unacceptable and abnormal response is now viewed as the "new normal" or even an "acceptable abnormal." As such, a more systems-oriented approach is needed for WFD within the typical HRM programs. That is, precisely the premise of this chapter. The long duration and high intensity of the lockdown act together to compound the landscape of HRM. The pandemic is a worldwide concern and must be addressed by a world-system viewpoint. Now, rather than later, is the time to start developing a new system of WFD to get ahead of the current pandemic and be prepared for future disruptions that may be caused by other pandemics or adverse world events. The COVID-19 global pandemic has changed our perspective on a lot of organizational pursuits, including WFD and HRM. Organizations, whether public or private, are grappling with what to do in managing human talent versus creating experience through WFD. Since WFD translates to economic development, citizens of any workforce deserve better developmental support, regardless of race, creed, gender, and ethnicity. Recent dramatic changes brought on by the pandemic include the following:

- We now have a masked workforce.
- Working at home is now the norm rather than the exception.
- It takes extraordinary self-discipline to work from home.
- Stability in the workforce has been severely impacted.
- Handshake deals are a thing of the past.
- Social distancing has emerged as being in vogue to replace social embracing.

- Strangers are even stranger now.
- Uncertainty is certain in the economy, workplace infrastructure, schools, businesses, industries, and governments.
- Security, safety, and social order are threatened.
- The normal upscale learning curves of human resources have been upended to be replaced by a call to "flatten the curve" (of COVID-19).
- Reluctance reigns in HRM.

On top of all of these COVID-19-caused changes, social and racial mass protests have disrupted even the best workforce plans. We need an enhanced approach to HRM and WFD. The workforce, once developed, must be managed with an alternate systems-informed methodology. This is what this chapter offers. The global-focused workforce of today must work collaboratively, creatively, culturally, and innovatively in a diverse and ever-changing work environment. It is only through a composite systems-based methodology that this global aspiration can be accomplished.

Due to COVID-19 pandemic disruption, the global workforce has been drastically affected and will continue to be changed, potentially forever. Our approaches and understanding of WFD, therefore, must adapt to the "new normal." Past approaches and practices of WFD may be the best methodology, henceforth.

LESSONS FROM THE LITERATURE

Holzer and Nightingale (2007) present various views and strategies for reshaping the American workforce in a changing economy. Now, we need to do research and development in reshaping worldwide workforce in the wake of COVID-19 pandemic. WFD and HRM are common challenges faced by all organizations, whether in public or private sector. This is evidenced by the wide body of knowledge available in the literature on the topic. Since the dawn of history, the management of human resources has been a challenge that has been addressed by a variety of techniques, ranging from autocratic systems and communal systems to democratic systems. See Barlow and Rochon (2019); Blomme (2010); Foegen (1980); Kim and Choi (2010); Lee and Sanders (2013); Maslow (1943); Pichler et al. (2016); Ruokolainen (2018); Soltis et al. (2013).

The WFD targets the objective of channeling the inherent capabilities of human resources in the path of advancement to satisfy organizational needs. Using three case studies in Belgium, Pichault (2007) presents human resource management-based reforms in public organizations. In his methodology, he introduced an analytical framework to compare and contrast public-oriented programs for HRM. Edwards, Colling and Ferner (2007) address the transfer of employment as a foundational approach for WFD in multinational companies. The converse to the public platform presented by Pichault (2007) is the private-equity view of HRM presented by Clark (2007). Schuler, Dowling and De Cieri (1993) present an integrative framework for international HRM. The framework could be effective for dealing with the type of disruption caused by COVID-19. Barlow and Rochon (2019) present a model for workforce-driven economic development, using the aerospace and defense industry in Southwest Ohio in the United States.

Prior to COVID-19, WFD concerns often revolve around the aging workforce, recruitment, training, retention, and learning processes. One dimension of WFD that was not previously addressed is workforce preservation. The topic of innovations in human resource preservation (post WFD) was first cogently addressed by Badiru and Barlow (2020a, 2020b). They assert that we typically focus on technical tools as the embodiment of innovation. But more often than not, process innovations might be just as vital. WFD, in particular, is more process development than tool development. There are numerous human factor strategies that can enhance the outcomes of WFD. Some of the innovations they recommend include paying attention to the hierarchy of needs of the worker (primarily safety and health in a COVID environment), recognizing the benefits of diversity, elevating the visibility of equity, instituting efforts to negate adverse aspects of cultural bias, and appreciating the dichotomy of socioeconomic infrastructure. While not too expensive to implement, these innovative strategies can be tremendously effective, particularly if extended to a global systems methodology, as presented in this chapter. Flynn (2014) presents a linkage of worker safety to the demographics of the workforce. There is WFD on one side, workforce redevelopment on another side, and workforce preservation on a third side. Workforce redevelopment is a topic not very often discussed, but COVID-19 brings its importance to the forefront. Redevelopment will be needed not only to boost the quantity of the productive capacity but also to restore and augment the capability, availability, and reliability of the workforce beyond the previous performance yardstick. Blustein (2006) addresses the psychology of working in the context that relates to how a worker might receive, accept, and leverage the available WFD opportunities. Preservation of a well-developed workforce can only be assured through innovative health and safety safeguards, as well as new organizational processes and procedures that will take a thorough systems-based understanding of the recurring risks that may be posed by a pandemic.

CONCEPT OF WORKFORCE PRESERVATION

Not only should we develop and preserve the workforce, we must also ensure and advance the quality of life and standard of living for the workforce. We often think of WFD as something that is done *a priori* to the actual job engagement. But an expanded systems-oriented approach has a wider spectrum of continuum, ranging from educational preparation, experiential preparation, recruitment practices, hiring processes, onboarding, training, retention, career advancement, and human resource preservation. Post COVID-19, the work environment will never go back to where things were before the pandemic.

Thus, now is the time to start instituting new WFD strategies and processes from a systems perspective. Even before the COVID-19 pandemic, the workforce has become increasingly more mobile with remote access and teleworking arrangements. The gradual decline of face-to-face work interfaces, now fueled by COVID-19, has created more challenges in the organizational communication and coordination infrastructure. Managing virtual teams in the workforce requires a new approach to workforce management. It is believed that COVID-19 will have a lasting impact. So, any new approached developed must have a lasting systems-based footprint.

If we preserve the health, welfare, and well-being of the workforce, we can sustain the contributions and productivity of the workforce. The concept of Total Worker Health[®], presented by Schill and Chosewood (2013), Schill (2016), and Schulte et al. (2015), is intertwined with this chapter's idea of workforce preservation. Work, wellness, and wealth can go hand in hand. Workers' health directly affects a nation's gross domestic product, based on a systems view of work, cascading from one person's level all the way (collectively) to the national level. Budd and Spencer (2014) also address the importance of linking workforce well-being to the content of work. Good health is related to good work performance. Health is an individual attribute that compliments each worker's hierarchy of needs. Without good health, even the best worker cannot perform. Without good health, even the best athlete cannot succeed. Without good health, even the most proficient expert cannot deliver his or her expertise. Without total health, even the most dedicated and experienced employee cannot contribute to the accomplishment of the organization's mission. Good health is a key part of the systems view of the workforce, as opined in this chapter. Chida and Steptoe (2008), Cherniack (2013), and Hemp (2004) also present ties between workforce well-being and availability to contribute to organizational work. This chapter's view is that a developed workforce that is not available to do the work adversely affects overall workforce preservation.

SYSTEMS VIEW OF WFD

What is a system? In a conventional definition, a system is defined as a collection of interrelated elements, whose collective and composite output, together, is higher than the mere addition of the outputs of the individual elements. With this viewpoint, each element in the system is recognized as a key driver or cornerstone in the overall system. Systems integration makes the world run smoothly for everyone. When something flares up in one corner, it can quickly spread to other parts, as we have seen in COVID-19 and the recent social unrest and protests. Systems thinking enhances the social conversation as we endeavor to provide, not just scholarly insights into world developments, but also social appreciation of what the workforce needs in terms of education, mentoring, culture, diversity, work climate, gender equity, job training, leadership, respect, appreciation, recognition, reward, work compensation, digital work environment, career advancement opportunities, hierarchy of needs, and other dimensions of the human environment.

There are several moving parts in WFD. Only a systems view can ensure that all components are factored into the overall methodology. A systems view of the world allows an integrated design, analysis, and implementation of WFD strategies. It would not work to have one segment of the enterprise embarking on one strategic approach while another segment embraces practices that impede the overall achievement of an integrated workforce. In the context of operating in the global environment, whether a process is repeatable or not, in a statistical sense, is an issue of workforce stability and sustainment. A systems-based framework allows us to plan for prudent utilization of scarce human resources across all operations, particularly post the COVI-19 pandemic. Sustaining the workforce through and beyond the pandemic is essential for the stability of operations. Some specific areas of workforce sustainment include the following:

- Environmental comfort in the workplace
- Operational sustainment for the organization
- Health sustainment for the workforce
- Welfare sustainment of the workforce
- Safety sustainment for workforce preservation

- Security sustainment in the workplace
- Work sustainment for workforce retention
- Financial sustainment for workforce compensation
- Economic sustainment for workforce's role in economic development
- Social sustainment for workforce connectivity during and after the pandemic

Because of COVID-19 pandemic and the coincidental social unrests, there is a growing worldwide call for equity and inclusion. This begs for looking at human interactions from a systems perspective for caring about and watching out for one another. It is through a systems viewpoint that we can advance the promotion of workforce diversity, inclusion, and tolerance.

WFD WITHIN THE HIERARCHY OF NEEDS

The psychology theory of "Hierarchy of Needs," postulated by Abraham Maslow in his 1943 paper, "A Theory of Human Motivation" (Maslow 1943), still governs how we respond along the dimensions of developing and managing human resources, particularly where group dynamics and organizational needs might have been disrupted by COVID-19. A socially induced disparity in the hierarchy of needs implies that we may not be able to fulfill personal and organizational responsibilities along the spectrum of a stable workforce. In a diverse workforce, the specific levels and structure of needs may be altered from the typical steps suggested by Maslow's Hierarchy of Needs. This calls for evaluating the needs from a multidimensional systems-based perspective. For example, a three-dimensional view of the hierarchy of needs can be used to coordinate personal needs with organizational needs with the objective of facilitating better WFD. Maslow's Hierarchy of Needs consists of five stages, which are all essential for a structural pursue of WFD in any organization. The needs are broadly summarized as follows:

1. **Physiological needs:** These are the needs for the basic necessities of life, such as food, water, housing, and clothing (i.e., survival needs). This is the level where access to money is most critical. The workforce at this level of need is primarily motivated by income generation to support the basic necessities of life.
2. **Safety needs:** These are the needs for security, stability, and freedom from physical harm (i.e., desire for a safe environment).

At this stage, the workforce is concerned about the assurance of safety, security, and stability in the workplace.

3. **Social needs:** These are the needs for social approval, friends, love, affection, and association (i.e., desire to belong). For example, social belonging may bring about better economic outlook that may enable each individual to be in a better position to meet his or her social needs. At this stage, issues such as fairness, inclusion, equity, acceptance, racial harmony, and bias-free work environment take on more importance.

4. **Esteem needs:** These are the needs for accomplishment, respect, recognition, attention, and appreciation (i.e., desire to be known). At this level of the hierarchy of needs of the workforce, equality of treatment, appreciation, and veneration become important.

5. **Self-actualization needs:** These are the needs for self-fulfillment and self-improvement (i.e., Desire to arrive). This represents the stage of opportunity to grow professionally and be in a position to selflessly help others. At this level, the individual in the workforce is celebrated and contented with accomplishments. The worker may have a sense of "having arrived."

Figure 8.1 presents a three-dimensional view adaptation of Maslow's Hierarchy of Needs to incorporate organizational WFD needs. The figure expands the hierarchy of needs to generate a three-dimensional rendition that incorporates organizational workforce-development hierarchy of needs. The location of each organization along its hierarchy of needs will determine how the organization perceives and embraces WFD programs. Likewise, the hierarchy position of each individual will determine how he or she practices commitment to personal development pursuits. Ultimately, the need for and commitment to WFD boil down to each worker's perception based on his or her location on the hierarchy of needs and the level of awareness of organizational WFD. How do we explain to a hungry poor family in an economically depressed part of the world, under COVID-19 pandemic, the need to conserve forestry? Or, how do we dissuade an old-fashioned professor (pre-COVID-19) from the practice of making volumes of hardcopy handouts instead of using electronic distribution? Cutting down on printed materials is an issue of advancing sustainment in a constrained-resource environment. In each wasteful eye, "the *need* erroneously justifies the *means.*"

In an economically underserved culture, most workers will be at the basic level of physiological needs; and there may be constraints on moving from one level to the next higher level. This fact has an implication on how human interfaces impinge upon organizational needs. In terms of organizational hierarchy of needs, the levels in Figure 8.1 are characterized as follows:

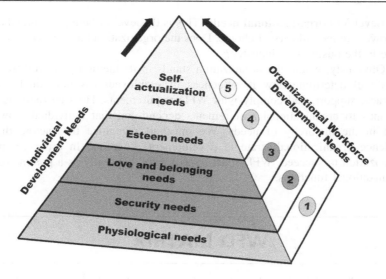

FIGURE 8.1 Three-Dimensional Adaptation of Maslow's Hierarchy of Needs to Incorporate Organizational WFD Needs

Level 1 of organizational needs: This is the organizational need for basic essentials of economic vitality to support the generation of value for stockholders and employees. Can the organization afford recovery programs following COVID-19 from cash reserves?

Level 2 of organizational needs: This is a need for organizational defense. Can the organization feel safe from external attack, such as mass protests or cyberattacks? Can the organization protect itself from hostile takeover attempts?

Level 3 of organizational needs: This is the need for an organization to belong to some market alliances and seen as a market leader, on the basis of its capable workforce. Can the organization be invited to join trade groups and participate in setting industry standards? Does the organization have a presence on some world stage?

Level 4 of organizational needs: This is the level of having market respect and credibility that can help preserve its image following a disruption of its productive infrastructure. Is the organization esteemed in some aspect of market, economic, or technology movement in the post-COVID-19 recovery efforts? What positive and stable accomplishment is the organization known for?

Level 5 of organizational needs: This is the level of being acclaimed as a "Power" in the industry of choice. Does the organization have a recognized niche in the business or industry?

Obviously, where the organization stands in its hierarchy of workforce goals will determine how it influences its employees (as individuals) to embrace, support, and comply with WFD requirements. How an individual responds to organizational requirements depends on that individual's own level in the hierarchy of needs. We must all recognize the factors that influence organizational strategic development programs. In order for an organization to succeed, WFD must be expressed explicitly as a goal across organizational functions.

WFD MATRIX

The coupling of technical assets and managerial tools are essential for achieving WFD. This chapter introduces a simple tool for organizing the relevant factors associated with WFD. The tool, called *WFD matrix*, overlays sustainment awareness factors, technical assets, and managerial tools. Figure 8.2 shows the matrix framework, while Table 8.1 shows the generic matrix design, considering both technical and managerial factors. The sample elements illustrate the nature and span of factors associated with WFD programs. Each organization must assess its own environment and include relevant factors and issues within the context of prevailing workforce environment that has been disrupted by COVID-19. Without a rigorous analytical framework, development pursuits will just be in words rather than deeds. One viable strategy is to build collaborative STEM (Science, Technology, Engineering, and Mathematics) alliances across organizations. The resiliency demonstrated by one collaborating organization can positively influence the response and operation of another collaborating organization. The analytical framework of systems engineering provides a tool for this purpose from an interorganizational perspective. With this, physical infra-structure, production systems, and human interface systems can be sustainably tied together to provide win–win benefits for all. An effective collaborative structure would include researchers and practitioners from a wide variety of disciplines, from both technical and humanities areas. Figure 8.2 represents the diverse elements involved in the WFD matrix. It recognizes both the technical and human aspects of managing human resources. Of particular interest is the role of direct communication, implicit cooperation, and explicit coordination, which form the triangular factors in

TABLE 8.1 Systems Framework for WFD Matrix

Managerial Environmental Factors

Technical Factors	Organizational behavior	Personnel culture	Resource baseline	Organization influence	Diversity and equity
Physical Infrastructure	Communication Modes	Cooperation Incentives	Coordination Techniques	Performance Management	Economics of participation
Work Design	Technical assignments	Work measures	Project design	Financial implications	Work control
Analytical Modeling	Resource combinations	Qualitative risk	Workforce productivity	Value-added assessment	Forecast models
Scientific Limitation	Work efficiency	Technical workforce	Contingency planning	Promotion potential	Educational support
Technology Constraints	Workforce preservation	Training programs	Operational risk management	Public acceptance	Technology risks

FIGURE 8.2   Factors Inherent in Post-COVID-19 WFD Programs

the Triple C model introduced by Badiru (2008). Table 8.2 presents systems framework for the WFD matrix within the context of HRM. The idea of the table is to enumerate the diverse technical factors vis-à-vis the managerial environmental factors. This expands the scope of considerations beyond the typical focus on only the management factors driving WFD. Essentially, the creation of WFD strategies must consider not only the organizational needs but also the personal hierarchy of needs of the individuals. With this expanded approach, we believe that WFD programs will be more robust and more sustainable.

SYSTEMS-BASED METHODOLOGY FOR WFD

Now that the various dimensions of HRM within the context of WFD have been addressed, we now put the elements within a specific systems modeling framework by using the DEJI systems model (Badiru 2014, 2019), which has stages for design (of a process), evaluation (of the process), justification (for adopting the process), and integration (of the process into normal business). The efficacy of the model centers around its focus on strategy integration. No matter how good a strategy is, it may fail in implementation if it is not integrated into the normal business process of an organization. The DEJI systems model provides a hierarchical framework for process design, evaluation, justification, and integration, such that most of the pertinent factors of WFD are considered both at the outset and at the implementation stages. Because of its structured stages, the DEJI systems model provides a continuous integration of WFD best practices into organizational strategies.

TABLE 8.2 DEJI Systems Model Matrix Application for HRM and WFD

DEJI Systems Model Components	HRM Description	WFD Techniques
Design	• Define HRM goals • Set human expectations • Identify critical factors	• Work standards • Performance metrics • Critical factors
Evaluate	• Set measurement parameters • Assess attributes • Evaluate benchmarks	• Benefit–Cost ratio • Workforce retention • Retirement benefit
Justify	• Economic development impacts • Technical and managerial feasibility • Alignment with strategic goals	• Education benefits • Workforce equity • Legal requirements
Integrate	• Identify common elements • Verify symbiosis • Check value synergy	• Workforce preferences • Hierarchy of needs • Teamwork synergy

A good WFD program should be at the intersection of efficiency, effectiveness, and productivity. Efficiency provides the framework for program robustness in terms of resources and the inputs required to achieve the desired level of organizational and personal satisfaction. Effectiveness refers to the implementation of program elements to meet specific needs and requirements of the organization. Productivity is an essential factor in the pursuit of workforce output as it relates to the throughput of the organization's mission. To achieve the desired levels of quality, efficiency, effectiveness, and productivity, a systems framework must be adopted. The DEJI systems model was developed for an integrative product design and development application, but it is applicable to the challenges of human WFD. The elements of the model pertinent for WFD are explained in the following:

- Design in the DEJI systems model embodies agility in defining the end goal and engaging both the internal and external stakeholders.
- Evaluation in the DEJI systems model embodies a realistic assessment of the feasibility of the WFD goal and gathering evidence of the effective metrics desired.
- Justification in the DEJI systems model embodies confirming the desirability of the intended outcomes and an articulation of the buy-in of the stakeholders.
- Integration in the DEJI systems model embodies the critical elements of a sustainable implementation, practicality of the implementation, the

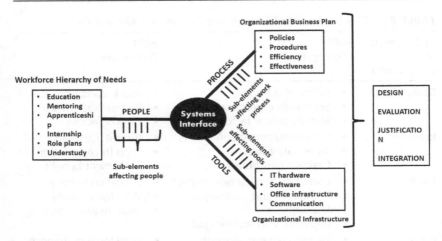

FIGURE 8.3　Input–Output DEJI Systems Model Framework for WFD

> affordability to the organization's business plan, and, mostly importantly, the alignment with personal preferences and desires of the workforce.

Most organizational failures occur when well-designed programs do not align with how the workforce actually works. This aspect of the model is what makes the DEJI systems model unique from other systems engineering models, such as the V-model, waterfall model, spiral model, walking skeleton model, and so on (Badiru 2019). For application purposes, the elements of the DEJI systems model interface and interact systematically to enhance the overall operational performance of organizational pursuits. Product quality management in industry is one fruitful application of the model (Badiru 2014). Under an implementation of the model, the design of work itself can be blueprinted to enhance the associated WFD to do the work (Badiru and Bommer 2017).

Figure 8.3 places all the aforementioned elements together into a guiding flowchart of organizational actions. This framework ensures that at least most of the critical elements of an organization's WFD strategy are addressed structurally.

CONCLUSIONS

A systems approach facilitates building a resilient workforce, upon which economic development is built. This chapter has presented an argument for using

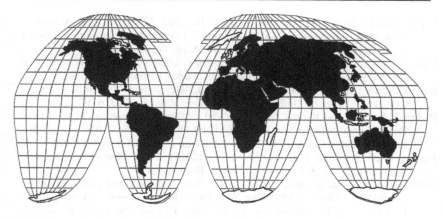

FIGURE 8.4   Fluidity and Cross-Hatching of World WFD

a systems approach for developing an organization's strategy for WFD. Rhetoric and practices that further entrench workforce discrimination are forestalled if we adopt a systems view of workforce interactions. The proposed systems methodology improves the ease of human resources compliance, highlights the common knowledge of what the law requires, paves the way for a consistency of HRM operations, ensures a standardized interpretation of policies, clarifies procedures, serves as a template for uniform policies, and serves as a guide for procedural remedies. The DEJI systems model utilized in the methodology of this chapter facilitates a structural design, evaluation, justification, and integration of WFD processes, policies, and strategies. Further, the proposed systems structure creates a pathway for workforce training and management situational awareness. We envision that the contents of this chapter will trigger additional interests in research studies on how post-COVID-19 WFD strategies can be planned, structured, executed, and sustained. Most organizations will not return to pre-COVID-19 normal. Consequently, going forward, things have to be done differently, the systems way. Figure 8.4 demonstrates the fluidity and cross-hatching interfaces of WFD across our present world.

REFERENCES

Badiru, Adedeji B. (2008), *Triple C Model of Project Management: Communication, Cooperation, and Coordination*, Taylor & Francis CRC Press, Boca Raton, FL.

Badiru, Adedeji B. (2014), "Quality Insights: The DEJI Model for Quality Design, Evaluation, Justification, and Integration," *International Journal of Quality Engineering and Technology*, Vol. 4, No. 4, pp. 369–378.

Badiru, Adedeji B. (2019), *Systems Engineering Models: Theory, Methods, and Applications*, Taylor & Francis/CRC Press, Boca Raton, FL.

Badiru, Adedeji B. and Barlow, Cassie (2020a), "Developing Workforce in Era of COVID-19," *Dayton Daily News Newspaper*, May 15, 2020, p. B7.

Badiru, Adedeji B. and Barlow, Cassie (2020b), "A Systems Approach to Workforce Development Amidst and Post COVID-19 Pandemic," *Working Paper*, Southwest Ohio Council for Higher Education (SOCHE), Dayton, Ohio, USA.

Badiru, Adedeji B. and Bommer, Sharon C. (2017), *Work Design: A Systematic Approach*, Taylor & Francis/CRC Press, Boca Raton, FL.

Barlow, Cassie B. and Rochon, Kristy (2019), "The Aerospace and Defense Industry in Southwest Ohio: A Model for Workforce-Driven Economic Development." In Badiru, A. B. and Barlow, C. B. (eds), *Defense Innovation Handbook: Guidelines, Strategies, and Techniques*, CRC Press/Taylor & Francis, Boca Raton, FL.

Blomme, R. J., Rheedeb A., and Tromp, D. M. (2010), "The Use of the Psychological Contract to Explain Turnover Intentions in the Hospitality Industry: A Research Study on the Impact of Gender on the Turnover Intentions of Highly Educated Employees," *The International Journal of Human Resource Management*, Vol. 21, No. 1, pp. 144–162. DOI: 10.1080/09585190903466954.

Blustein, David L. (2006), *The Psychology of Working: A New Perspective for Career Development, Counseling, and Public Policy (Lea Series in Counseling and Psychotherapy)*, Lawrence Erlbaum Publishers, Mahwah, NJ.

Budd, J. W. and Spencer, D. A. (2014), "Worker Well-Being and the Importance of Work: Bridging the Gap," *European Journal of Industrial Relations*, Vol. 1, No. 16, pp. 12–25.

Chida Y. and Steptoe, A. (2008), "Positive Psychological Well-Being and Mortality: A Quantitative Review of Prospective Observational Studies," *Psychosomatic Medicine*, Vol.70, No. 7, pp. 741–756.

Cherniack, M. (2013), "Integrated Health Programs, Health Outcomes, and Return on Investment: Measuring Workplace Health Promotion and Integrated Program Effectiveness," *Journal of Occupational Environmental Medicine*, Vol. 55, No. 12 supplement, pp. S38–S45.

Clark, Ian (2007), "Private Equity and HRM in the British Business System," *Human Resource Management Journal*, Vol. 17, No. 3, pp. 218–226.

Edwards, Tony, Colling, Trevor and Ferner, Anthony (2007), "Conceptual Approaches to the Transfer of Employment Practices in Multinational Companies: An Integrated Approach," *Human Resource Management Journal*, Vol. 17, No. 3, pp. 201–217.

Flynn, M. A. (2014), "Safety and the Diverse Workforce," *Professional Safety*, Vol. 14, No. 59, pp. 52–57.

Foegen, J. H. (1980), "Time-Off without Pay: Value Can Exceed Cost," *Human Resource Management*, 20, 10–12. doi 10.1002/hrm.393020020.

Hemp P. (2004), "Presenteeism: At Work – but Out of It," *Harvard Business Review*, Vol, 82, No. 10, pp. 49–58.

Holzer, Harry J. and Nightingale, Demetra S. (eds) (2007), *Reshaping the American Workforce in a Changing Economy*, The Urban Institute, Washington, DC.

Kim, M. S., and Choi, J. N. (2010), "Layoff Victim's Employment Relationship with a New Employer in Korea: Effects of Unmet Tenure Expectations on Trust and Psychological Contract," *The International Journal of Human Resource Management*, Vol. 21, No. 5, 781–798.

Lee, S. and Sanders, R. M. (2013), "Fridays Are Furlough Days: The Impact of Furlough Policy and Strategies for Human Resource Management During a Severe Economic Recession," *Review of Public Personnel Administration*, Vol. 33, No. 3, 299–311. https://doi.org/10.1177/0734371x13477426.

Maslow, Abraham H. (1943), "Theory of Human Motivation," *Psychological Review*, Vol. 50, No. 4, pp. 370–396.

Pichault, Francois (2007), "HRM-Based Reforms in Public Organizations: Problems and Perspectives," *Human Resource Management Journal*, Vol. 17, No. 3, pp. 265–282.

Pichler, S., Varma, A., Michel, J. S., Levy, P. E., Budhwar, P. S., and Sharma, A. (2016), "Leader Member Exchange, Group- and Individual-Level Procedural Justice and Reactions to Performance Appraisals," *Human Resource Management*, Vol. 55, No. 5, pp. 871–833. https://doi.org/10.1002/hrm.21724.

Ruokolainen, M., Mauno, S., Diehl, M. R., Tolvanen, A., Mäkikangas, A., and Kinnunen, U. (2018), "Patterns of Psychological Contract and Their Relationships to Employee Well-Being and In-Role Performance at Work: Longitudinal Evidence from University Employees," *The International Journal of Human Resource Management*, Vol. 29, No. 19, 2827–2850. https://doi.org/10.1080/09585192.2016.1166387.

Schill, Anita L. (2016), "Advancing Well-being through Total Worker Health," Keynote Address, 17th Annual 2016 Pilot Research Project (PRP) Symposium, University of Cincinnati, Cincinnati, OH, October 13.

Schill, Anita L. and Chosewood, L. C. (2013), "The NIOSH Total Worker Health Program: An Overview," *Journal of Occupational Environmental Medicine*, Vol. 55, No. 12 supplement, pp. S8–S11.

Schuler, R., Dowling, P. and De Cieri, H. (1993), "An Integrative Framework of Strategic International Human Resource Management," *International Journal of Human Resource Management*, Vol. 4, No. 4, pp. 717–764.

Schulte, Paul A. et al. (2015), "Considerations for Incorporating 'Well-Being' in Public Policy for Workers and Workplaces," *American Journal of Public Health*, Vol. 105, No. 8, August, pp. e31–e44.

Soltis, S. M., Agneessens, F., Sasovova, Z., and Labianca, G. J. (2013), "A Social Network Perspective on Turnover Intentions: The Role of Distributive Justice and Social Support," *Human Resource Management*, Vol. 52, No. 4, pp. 561–584. https://doi.org/10.1002/hrm.21542.

Sustainability Case Study of End-of-Life Vehicles: Generalized Framework for the Design of Eco-Industrial Parks

9

"Lesson learned should be lesson practiced."

— *Adedeji Badiru*

Adapted(not reprinted) with Open-Access Author's consent from

Al-Quradaghi, Shimaa, Zheng, Quipeng P., and Eklamel, Ali (2020), "Generalized Framework for the Design of Eco-Industrial Parks: Case Study of End-of-Life Vehicles," *Sustainability*, Vol. 12, p. 6612. doi:10.3390/su12166612

BACKGROUND

Al-Quradaghi et al. (2020) present an excellent example of how industrial development pursuits can take cognizant of the global push for sustainability. This chapter presents a condensed synopsis of the case study project published by the authors. Without a loss of generality and due to book-chapter space limitation, only the key conceptual framework of the case study is presented here without the figures and artwork of the study report. Interested readers are directed to Al-Quradaghi et al. (2020) for the full details.

Eco-industrial parks (EIPs) are promoting a shift from the traditional linear model to the circular model, where industrial symbiosis plays an important role in encouraging the exchange of materials, energy, and waste. This chapter proposes a generalized framework to design EIPs and illustrates it with regard to the end-of-life vehicle (ELV) problem. An EIP for ELVs (EIP-4-ELVs) creates a synergy in the network that leverages waste reduction and efficiently uses resources. The performance of the proposed framework is investigated along with the interactions between nodes. The proposed framework consists of five steps: (1) finding motivation for EIP, (2) identifying all entities with industrial symbiosis, (3) pinpointing the anchor entity, (4) determining industrial symbiosis between at least three entities and two exchange flows, and (5) defining exchange-flow types. The last two steps are connected by a feedback loop to allow future exchange flows. The proposed framework serves as a guideline for decision makers during the first stages of developing EIPs. Furthermore, the framework can be linked to car-design software to predict the recyclability of vehicle components and aid in manufacturing vehicles optimized for recycling.

INTRODUCTION

An industrial ecosystem optimizes the consumption of energy and materials and minimizes the generation of waste (Frosch and Gallopoulos 1989). The study of industrial systems that operate like natural ecosystems is called industrial ecology, in which the natural ecosystem generates waste from one organism to be the resource for another (Frosch 1994). Similar to the natural ecosystem, the industrial ecosystem capitalizes on the exchanges of one firm's waste to be another firm's resource.

Industrial symbiosis is a subset of industrial ecology and has a particular focus on material and energy exchanges (Gu et al. 2013). EIPs develop when industrial symbiosis occurs between firms; the interactions include exchanges of material and energy.

EIPs promote a shift from the traditional linear model to the circular model. They are considered a community of businesses that reduce the global impact by sharing resources such as materials, energy, and water to reduce waste and pollution and increase economic gains (Chertow 2007). The interactions in the community improve the environmental performance of the industrial network. EIP is promoting a shift from the traditional linear model of "raw material to industry to waste" to a closed-loop model of "raw material to industry A to waste to raw material to industry B." Considering the global strategy, cities are considered at macro level, and single industry at the micro level, and in between, there is the meso level where EIPs exist. EIPs not only exchange waste heat, steam, biowaste, and industrial waste but also knowledge, material, and energy. The circular process that is built within EIP connects the entities and results in minimum effect on the environment. Hence, solving environmental problems using an EIP approach is one effective strategy of waste management that limits pollution impacts on the environment. In this way, EIPs can be implemented within industrial districts to encourage ecodesign practices and the transition to circular business models that are more sustainable in nature. This will also improve the efficiencies of the existing industrial entities in given districts and thus enables them to improve their competitiveness on a global scale. The first known EIP was in Kalundborg, Denmark; industrial symbiosis gradually evolved over 20 years (Ehrenfeld and Gertler 1997). Today, many countries have EIP projects: Argentina, Austria, Brazil, Canada, China, Denmark, Finland, France, Germany, Italy, Netherlands, Norway, Singapore, South Korea, Spain, Sweden, Switzerland, the United Kingdom, the United States, and others (Aid et al. 2017; Behera et al. 2012; Chertow 2007; Gu et al. 2013; Mat et al. 2016; Piaszczyk 2011; Susur et al. 2019; Tessitore et al. 2015). Other works vis-a-vis EIPs covered multidisciplinary areas, for example, optimization (Boix et al. 2015), lifecycle assessments (Zhang et al. 2017), policy implementation (Jiao and Boons 2014), social networks (Song et al. 2018), and topology (Chertow 2007). Furthermore, several studies in the literature proposed frameworks for EIPs. For example, on December 2017, the UN Industrial Development Organization (UNIDO), joining efforts with German Development Co-Operation (GIZ) and the World Bank Group (WBG), published a document that presents an International Framework for Eco-Industrial Parks (UNIDO, GIZ 2017). The International Framework for Eco-Industrial Parks focuses on four performance categories: park management, environmental, social, and economic (see Al-Quradaghi et al. 2020). Each category consists of prerequisites and performance requirements that can be measured. All prerequisites and performance requirements must be met for something to be considered an EIP. The report guides decision makers on the important components to achieve maximal benefits economically, environmentally, and socially. However, as indicated in the report, the International Framework for EIPs provides only strategic details for EIP requirements and does not translate them to existing EIPs.

The International Framework for Eco-Industrial Parks created by UNIDO, GIZ, and WBG is not the only initiative to provide essential elements for forming EIPs. Several studies in the literature proposed frameworks for EIPs from different views. There are two common themes for these frameworks: general and special case. While researchers like Haskins (2007), Sopha et al. (2009), Boons et al. (2011), Dumoulin et al. (2016), Konstantinova et al. (2019), Andiappan et al. (2016), Yedla and Park (2017), Romero and Ruiz (2013), and Tao et al. (2019) proposed a general framework for EIPs, other researchers like Behera et al. (2012), Liu and Côté (2017), and Gopinath et al. (2018) proposed more specific frameworks for solving a special-case issue. The reviewed articles are presented in the following paragraphs in each theme.

Haskins (2007) proposed a general framework for EIP development named iFACE. The acronym refers to i – identify stakeholders and their needs; F – frame the problem(s); A – alternatives identification and study the options; C – choose and implement a course of action; and E – evaluate continuously. The author expressed the framework as a combination of system engineering (SE), industrial ecology, organizational dynamics, logistics, and supply-chain theories.

Sopha et al. (2009) presented a more extended framework for creating industrial-symbiosis modeling. The framework consists of two parts: (1) an SE process and (2) methods. The SE process consists of six steps: needs identification, define the requirements, specify the performances, analyze, design and improve, and implementation. The methods component lists different enabling technologies for each SE process step. Interviewing was proposed for steps 1–3, brainstorming for step 1, literature study for steps 2–6, survey for steps 2–4, field study for steps 1–3, and workshop for steps 3–6. The authors applied the framework on the case of an industrial site in Mongstad, Norway, to increase industrial symbiosis (Sopha et al. 2009).

Boons et al. (2011) proposed a conceptual framework for analyzing the dynamics of industrial symbiosis. The framework has a set of conditions that are referred to as "antecedents" that affects another set of "mechanisms." The mechanisms elaborate on two levels: (1) societal and (2) regional industrial systems. The application of the mechanisms led to outcomes that were reflected on the ecological system and social networks. The authors concluded that the framework helped to build a theoretical understanding of the dynamics of industrial symbiosis (Boons et al. 2011).

Romero and Ruiz (2013) proposed a nested-system framework for modeling EIP operations. The framework describes the relationship between industrial systems and the environment. The main building blocks for the nested framework are economic, social, and natural systems. In applying the framework, five key properties were taken into consideration: (1) functionality; (2) theoretical knowledge; (3) adaptability; (4) reliability; and (5) life span. The authors supported the framework by merging complex-adaptive-system theory, industrial ecology, and the analysis of existing EIPs (Romero and Ruiz 2013).

Dumoulin et al. (2016) proposed an environmental-assessment framework for facilitated regional industrial symbiosis. The framework helps to identify all environmental impacts in facilitated regional industrial symbiosis. The framework was divided into two main sections: (1) logical basis, where the key elements of environment observation are identified; and (2) method, where three steps are performed, namely, identifying environmental impact, designing indicators, and assessing the environment. The authors applied the framework on a case in Réunion, a French territory in the Indian Ocean that had the potential for industrial symbiosis (Dumoulin et al. 2016).

Kuznetsova et al. (2016) discussed the challenges that EIPs face and proposed an optimization framework for EIP topology and operation. The framework consisted of two stages: optimization of EIPs' (1) topology and (2) operation. Each stage includes several steps. The framework considers uncertainties in EIP and provides appropriate predictions. The authors detailed the uncertainties and risks that should be taken into consideration at the design stage (Kuznetsova et al. 2016).

Andiappan et al. (2016) proposed an optimization-based framework for coalitions in EIP. The framework starts with defining cooperative plants in the intended EIP. Then, the framework continues identifying interactions between plants, uses mathematical models to calculate the symbiosis costs (cost of sharing exchanges between plants), and ends up with the EIP configuration. The authors used mathematical models for calculating the economic correlations of cost and savings allocation and performed stability analysis for each entity. The framework was applied to a palm-oil EIP in Malaysia. Results showed an increase in the savings for industries in the EIP (Andiappan et al. 2016).

Konstantinova, Johannes, and Vejrum (2019) discussed the importance of trust between stakeholders in industrial-symbiosis initiatives. They developed a conceptual industrial-symbiosis trust framework to facilitate gaining trust between partners. The framework illustrates three notions of trust (ability, integrity, and benevolence) through three trust bases: (1) calculus-, (2) knowledge-, and (3) identification-based trusts. In their paper, the authors answered the research question of "how can firms develop trust in the context of industrial symbioses investment" through proposing the framework that merges industrial symbiosis and management techniques (Konstantinova et al. 2019).

Tao et al. (2019) proposed a three-dimensional framework for studying the influence of policy on industrial symbiosis from the firm's perspective. The three dimensions are industrial-symbiosis (1) fostering models, (2) implementation stages, and (3) policy instruments. The framework was demonstrated on a horizontal axis that presented ten executive policy instruments, the vertical axis presented five stages of industrial-symbiosis implementation, and the depth axis presented four models of industrial-symbiosis fostering. The authors applied the framework on two existing EIPs, one in the United Kingdom and the other in China (Tao et al. 2019).

For special case frameworks, Behera et al. (2012) presented the existing research and development into business framework that was developed by the Ulsan EIP Center in South Korea. The framework consists of three main steps: (1) exploring new networks, (2) feasibility study, and (3) commercialization. Each step leads to another step following certain criteria within the framework. The aim of the framework was to design industrial symbiosis between EIP stockholders. As an example, the authors presented the Ulsan industrial symbiosis and explained how it had been developed to reach 40 instances of symbiosis, some of which having been designed (Behera et al. 2012).

Liu and Côté (2017) presented a framework for incorporating ecosystem services into China's EIPs. The framework combines policies, governance, techniques, technologies, key actors, and support organizations to build the industrial symbiosis. In the framework, two main components are the core: Component I, with elements of policies, governance, technologies, and business development; and Component II, with elements of key actors and support organizations. The framework suggested integration between the two components to result in an EIP. The framework was designed to solve China's environmental issues by proposing a circular economy through encouraging EIP development. The authors suggested that the framework could provide guidance for other EIPs around the world (Liu and Côté 2017).

Gopinath et al. (2018) presented a material-flow framework for the sugar industry on the basis of an extensive literature review. The authors reviewed the literature to find all characteristics of the sugar industry to identify the optimal route that resulted in minimizing waste. The framework details the material flow for the sugar industry and points to several waste types that can be reused by other industries. The chapter shows the importance of synergy between different types of sectors (Gopinath et al. 2018).

The earlier paragraphs provided an overview of some available frameworks for EIPs in the literature. The discussions outlined the efforts by scholars with different backgrounds to propose ways and methods for designing EIPs. The review elucidates that although considerable efforts have been done in the literature, essential research regarding applying the approach to different types of systems such as ELVs is still required. Furthermore, the literature revealed the complexity of the available frameworks. In the early stage of designing EIPs, decision makers need a simple, clear, and strategic framework to follow and depend on. After having a solid base about the foundation of the elements in EIPs with a general clear framework, decision makers and the EIP team might follow other comprehensive frameworks that pertain to specific needs, such as optimizing industrial-symbiosis flow, evaluating the firms' trust, proposing policies, and other areas of concern. There exists no generalized framework that compiles precise foundations and simple steps at the same time. Hence, there is a need to investigate and construct a generalized and simple-to-follow framework for the design of EIPs. This chapter bridges this gap in the literature. A simple and yet comprehensive framework is proposed in this chapter. The framework consists of five steps: (1) finding motivation for EIP, (2) identifying all entities with industrial symbiosis, (3) pinpointing the anchor entity, (4) determining industrial symbiosis between at least three entities and two exchange flows, and (5) defining exchange-flow types. Steps (4) and (5) are connected by a feedback loop, which allows any additional exchange flows in the future. The chapter illustrates the use of the framework on a special case study that involves ELVs. Applying the recyclability index proposed by Villalba et al. (2004), the framework can be linked to vehicle-design software to predict the recyclability of different types of components.

The remainder of this chapter is organized as follows. The following section presents the proposed framework for efficiently designing EIPs. In the third section, a case study is employed to illustrate the use of the framework. In the final section, the conclusions and future work are outlined.

GENERALIZED FRAMEWORK FOR EIP DESIGN

With the extensive approaches available in the literature, the need for a straightforward framework is rising. The proposed framework in this section illustrates the foundations of designing EIPs. It provides general step-by-step actions to be followed in the very early stages of designing EIPs. The framework answers fundamental questions that decision makers need in order to start developing and designing EIPs. In a simple, clear, and step-by-step strategy, the framework lays out necessary actors in the process of designing EIPs. The framework requires primary data that can easily be collected from each entity forming EIPs.

The generalized framework for the design of EIPs consists of five steps (see Al-Quradaghi et al. 2020): (1) finding motivation for EIP (Kuznetsova et al. 2016), (2) identifying all entities with industrial symbiosis (Chertow 2000), (3) pinpointing the anchor entity (Eilering and Vermeulen 2004), (4) determining industrial symbiosis between at least three entities and two exchange flows (Chertow 2007), and (5) defining exchange-flow types (Eilering and Vermeulen 2004). Steps (4) and (5) are connected by a feedback loop to allow additional exchange flows in the future. Used parts should first be put into a used market depending on historical demand. Information on their utilization should be monitored, and a decision to eventually move them to the recycling node and consider them as waste material should be made frequently. This balance between reuse of useful components and recycling them as waste is an important consideration in step 2 of the proposed methodology.

For creating a successful EIP, a clear motivation should be stated, whether environmental, economic, or mixed (Chertow 2000). From this point forward, identifying entities shapes the overall components of the network. Next, pointing out the main entity in the network plays a great role in outlining the anchor entity that attracts other entities toward it.

To ensure industrial symbiosis in the EIP, the 3–2 heuristic rule (Chertow 2007) should be satisfied. This rule states that "at least three different entities must be involved in exchanging at least two different resources" (Chertow 2007). Hence, it is necessary to generate a full list of possible exchanges in the network (Chertow 2007; Felicio et al. 2016; Heeres et al. 2004; Tian et al. 2014). For this reason, the EIP matrix is proposed to list all industrial symbioses in EIPs.

The last step is to identify material-exchange types. A figure presented in Al-Quradaghi et al. (2020) illustrates the different exchange types on the basis of the literature. In her study, to find the taxonomy of different material-exchange types Chertow (2000) classified five types: (1) through waste exchanges through third party; (2) within facility, firm, or organization; (3) between firms collocated in a defined EIP; (4) between local firms that are not collocated; and (5) between virtually organized firms across a broader region (Chertow 2000, 2007). However, she highlighted that "Types 3–5 offer approaches that can readily be identified as industrial symbiosis" (Chertow 2000). For that reason, we considered material-exchange types 3 to 5 to be the main types for flows to build EIP. On that basis, we developed an illustrative figure to simplify external-exchange types 3 to 5.

The framework has a feedback loop between steps (4) and (5) that allows for modification and addition whenever there is new industrial symbiosis in the system. The framework results in a connected graph with nodes and arcs. Nodes represent the entities in the network, and arcs represent the exchange flows (see Al-Quradaghi et al. 2020). The framework steps are elaborated in the next subsection.

Find Motivation

This step initially ensures the commitment of creating EIPs. The question, "Why create an EIP?" is very important, and on the basis of the answer, the motivation is clear. The motivation for forming EIP can be related to the three pillars of sustainability: economic, environmental, and social. Some companies exchange resources in order to reduce cost or increase profit (economic pillar). Other companies have industrial symbiosis as a way to reduce greenhouse-gas (GHG) emissions and waste (environmental pillar). To go beyond these two reasons, some companies form EIPs to create more job opportunities to people in the EIP region (social pillar). In fact, all these benefits can be met in EIP.

Identify Entities

After knowing the motivation behind forming an EIP, the next step is to identify all possible entities that help in achieving that motivation. Entities in the planned EIP should have exchange flows with other entities from which the EIP benefits.

Pinpoint Anchor Entity

From all entities listed in the previous step, there should be one entity that attracts other entities toward it as it has the most exchange flows to share. The anchor entity is the largest giver in the EIP. It is important to identify the anchor entity to ensure that industrial symbiosis between entities continues.

Determining Industrial Symbiosis

Exchange flows between entities should be listed and determined. This visualizes all possible industrial symbiosis in the EIP. For simplicity, we propose the EIP matrix that lists all industrial symbiosis in the EIP. The EIP matrix summarizes all exchange flows and gives detailed information about the network.

Define Exchange-Flow Types

Through a graphical presentation, Al-Quradaghi et al. (2020) highlight the different exchange types on the basis of the literature. The types of exchange flow can be defined in this step. In this chapter, we considered external-exchange types 3 to 5 as the main flow types to build an EIP. For industrial symbiosis to happen, there needs to be external and not internal exchange flow. We define "internal exchange" as any industrial symbiosis that exists within the entity (e.g., old equipment from one department in the entity can be used in another department). On the other hand, "external exchange" is any industrial symbiosis that appears beyond the boundary of the entity (e.g., old equipment sent to recycling company), including collocated, non-co-located, and regional firms. This requirement is very important, as it identifies the distance between entities and the modes of transportation of the exchange flows (trucks, pipelines, etc.).

FRAMEWORK APPLICATION

ELV Overview

The European Directive of ELVs 2000/53/EC defines ELVs as "vehicles that have become waste," and the waste defined as "any substance or object which the holder discards, or intends to discard, or is requires to discard" (European Parliament and Council 2000). According to the Official Journal of the European Communities, ELVs account for up to 10% of the total amount of waste generated annually in the European Union (European Parliament and Council 2000). The directive requires car manufacturers to ensure that a minimum of 95% by weight per vehicle is reusable and/or recyclable, including a minimum of 85% of material recoverability (recyclability) or reuse (Garcia et al. 2015).

The waste stream generated from ELVs can be controlled/regulated through the vehicle-design phase when product development occurs. "Design for X" is one of several methods to aid the designer in this phase, where "X" refers to the lifecycle phase being evaluated (Curlee et al. 1994; Meerkamm 2007; Tonnelier et al. 2007; Watson and Radcliffe 2010). In other words, "X" represents the aim of the design: recycling, quality, sustainability, cost, and so on. It requires the design to meet the specific defined goal (X).

One solution could be that of introducing design for sustainability (DfS) in the design phase that requires maximizing resource efficiency while minimizing environmental impact (Curlee et al. 1994; Reuter et al. 2013; Villalba et al. 2004). A more specific definition for DfS is given by Vezzoli et al. (2018): "A design practice, education, and research that, in one way or another, contributes to sustainable development."

Using recycled scraps benefits the environment threefold by: (1) saving raw materials, (2) saving energy, and (3) reducing GHG emissions (Andersson 2016; Reuter et al. 2013; Villalba et al. 2004). Taking steel as an example to demonstrate the use of recycling, every ton of new steel made from scrap steel saves 2500lb of iron ore, 1400lb of coal, and 120lb of limestone (Fold 1). The use of recycled scrap steel reduces energy use by 75% (Fold 2). The estimated GHG reduction for recycling steel in every recycled vehicle is 2205lb of GHGs (Fold 3; ARA (2012)). In general, energy used in recycling scrap materials is less than the energy used in manufacturing raw materials (Curlee et al. 1994; Graedel et al. 2011). Energy saving and GHG reduction (ARA 2012) for some metals are illustrated in the case study (see Al-Quradaghi et al. 2020).

However, the dilemma is to figure out if recycling ELVs is economically feasible. Villalba et al. (2004) proposed a recyclability index to measure "the ability of a material to regain its valued properties through recycling process" (Graedel et al. 2011). The recyclability index calculates the profit-to-loss margin for recycling; hence, it determines whether it is economically feasible to recover the material (Villalba et al. 2004). With a positive margin, recycling is a good choice; a negative margin indicates some concerns making the material not worth recycling.

The ELV recycling system aims to isolate hazardous content, and recover usable parts and recycle others (Andersson 2016). There is a tremendous number of studies in the literature about recycling systems and managing ELVs that were comprehensively studied by Simic (2013) and Karagoz et al. (2019).

Vehicles mainly go through different stages. Maudet et al. (2012) highlighted two main systems for treating ELVs: dismantling components and vehicle shredding. In more detail, Edwards, Bhamra, and Rahimifard (2006) described three main stages in the recycling process: (1) depollution, (2) dismantling, and (3) shredding. The case study report illustrates the overall stages for treating ELVs, including depollution, dismantling, and shredding (see Al-Quradaghi et al. 2020). The elements are summarized as follows:

- Depollution: Drain all fluids (gas, oil, coolant, etc.) and remove battery
- Dismantling: Remove engine, tires, windshield, and steering wheel
- Shredding: Press the hulk and send to shredding machine

In the first stage (depollution), all fluids are drained and the battery is removed. The second stage (dismantling) removes the engine, tires, wires, cables, windows, bumpers, and other parts that are useable. In the third stage (shredding), the vehicle's hulk is pressed using a hammer mill and then sent to the shredding machine. In the process, ferrous metal is separated using magnetic separation, and an eddy current is used to separate the nonferrous metal. The remainder of that process (plastics, rubber, fabrics, and dirt) is called automobile shredder residue (ASR) or fluff (Andersson 2016; Choi et al. 2005; Curlee et al. 1994; Edwards, Bhamra, and Rahimifard 2006; Paul 2009). According to Curlee et al. (1994), ASR generated from recycling ELVs accounts for about 25% (by weight) of the shredded material. The highest percentage of components in the ASR is fibers, which accounts for 42% by weight, followed by plastics with 19.3% by weight. There are several approaches to separate the components or use the ASR for different purposes (Baker et al. 1995; van Schaik and Reuter 2004).

ELV management is a crucial issue to deal with for governments, vehicle producers, and treatment facilities. It has received increased attention due to its implications, both economic and environmental. The problem has both tactical and strategic-level decision-making components. D'Adamo et al. (2020) prepared regression models to predict the amount of ELVs generated yearly as a function of GDP and population. They concluded that, given the great amount of ELVs generated, adopting a practical procedure for constructing efficient procedures to connect and induce collaborations between the actors involved in ELV will greatly help in enhancing sustainability and creating economic opportunities (D'Adamo et al. 2020). Karagoz et al. (2019) provided a comprehensive review of 232 peer-reviewed articles published in the period from 2000 to 2019 that was aimed at identifying the gaps in the ELV management literature. They concluded that only few researchers suggested solutions that closed the waste management loop by recycling and suggested that such approaches should be devised for the solution of real-life ELV management problems to generate reasonable solutions for them. Finally, in a recent article that provided a bibliometric literature review and assessed the efficiency of ELV management, de Almeida and Borsato (2019) concluded that the literature reveals a series of strategies that are confusing. The chapter outlined several ELV management strategies and the different processes involved. The chapter also concluded that waste management research focusing on the holistic nature of the ELV problem and which considers nodes of different efficiencies is still lacking. Furthermore, the chapter suggested future research management strategies that focus on sustainability and the triple bottom line. Therefore, the proposed strategy of the previous section is clearly a step forward toward bridging this gap and its use is illustrated on the case of the ELV management problem in the next sections.

The proposed methodology must also be tailored to the type of vehicles that are in existence in a certain country and also to the specifics of that country. For instance, Che et al. (2011) discuss the specifics of ELV problem in Japan, China, and Korea, and propose different scenarios. For example, labor cost is high in Japan and the design of an EIP (applied to the ELV management problem) must take this into account. This can be done by focusing on automation and taking the economics of the problem into consideration. For example, in the case of several possible alternatives available in one of the suggested nodes by the proposed methodology, the different scenarios must be compared based on a composite objective that considers both the Net Present Value (NPV) and the sustainability component. In this way, the most desirable alternative with respect to this composite objective is selected. The specific nature of a given country can also be in terms of enforced recycling laws. Step 2 of the methodology that focuses on the identification of all entities with industrial symbiosis must therefore be altered to consider only entities that conform to the law of the country where the recycling unit is to be implemented.

New generation vehicles (NGVs), such as hybrid, plug-in hybrid, and electric vehicles, are emerging into the market with increasing rates due to advances in battery technology, material design, and computerized technology. These vehicles have different components, compared to the traditional fossil fuel vehicles. They are equipped, for example, with highly efficient nickel-hydrogen or lithium-ion batteries. When the proposed methodology is applied to NGVs, in step 2, which is concerned with the identification of all entities with industrial symbiosis, the balance between reuse and recycling should be taken into account. For NGVs, there is an emerging trend for the effective utilization of waste batteries (Yu et al. 2017). Furthermore, because of the inclusion of these highly efficient batteries, the steel content of the vehicles is much lower than that of the traditional fossil fuel vehicles while plastic content is more. For these reasons, nontraditional processing and recycle nodes should be considered for the case of NGVs. Yu et al. (2017) provides a comprehensive analysis of the different recycle and reuse approaches of waste batteries from NGVs.

Eco-Industrial Park for End-of-Life Vehicles

The proposed framework was applied to solve ELVs (EIP-4-ELVs). The steps of the framework tailored to this case are illustrated in the case study report (see Al-Quradaghi et al. 2020). The motivations for the case of EIP-4-ELVs are environmental and economic. ELVs are harmful to the environment, so solving this problem is the main motivation. Furthermore, recycling old cars generates profit for many industries. The next step is to identify which industries form the EIP-4-ELVs. The entities are suggested to be power plants, dismantling facilities (DFs), waste-to-energy plants, wastewater-treatment plants, glass industries, tire recycling, aluminum, plastic, and steel companies, and battery-recycling or -refurbishing companies. The anchor entity that generates the most waste/by-products in the case of EIP-4-ELVs is the DF. This facility is the core of EIP-4-ELVs as it sends out scrap materials of different types to the corresponding industries. Next is to determine all possible exchange flows in the EIP-4-ELVs using the EIP matrix. Last is to define the exchange-flow type for transportation purposes – external type 5 in this case. The feedback loop between the last two steps allows for any future change or modification in the exchange flow. The EIP matrix is provided for this case in a tabulated format (see Al-Quradaghi et al. 2020). Cells with (–) in the table indicate possible industrial symbiosis in the future as the EIP evolves.

The anchor entity, as mentioned earlier, is the DF that is the main entity for sending by-products/waste to other industries. The waste and multiple by-products created from the DF are considered resources for the other industries. The proposed industrial symbiosis given by the EIP matrix is shown in a connected network presented in the study report (see Al-Quradaghi et al. 2020). Nodes present the industries in the EIP-4-ELVs, and arcs show the exchange flows. The network illustrates how entities in the proposed EIP-4-ELVs can utilize waste from each other. In the next section, the case study authors ran a simulation model to the developed network and present the outcomes. As the aim of this chapter is to solve ELVs, the focus was to study the material flows in the proposed system boundary.

EIP-4-ELV Simulation

Model Assumptions

The vehicle DF is the source for all types of vehicles. In this study, we based our calculations on data given by UN Environment Program (UNEP) in the 2013 report titled "Metal Recycling, Opportunities, Limits, Infrastructure" (Reuter et al. 2013), and from the original source, the European Commission Joint Research Center report, "Environmental Improvement of Passenger Cars". We considered two types of passenger vehicles: petrol and diesel. Al-Quradaghi et al. (2020) present the composition of an average passenger car from each type, showing the average curb weights for each type. As defined by the US Department of Transportation, curb weight is "the actual weight of the vehicle with a full tank of fuel and other fluids needed for travel, but no occupants or cargo" (Kahane 2003).

In the available data, this did not add up to the total weight because of the lack of detailed information (Kahane 2003). For the purpose of this chapter, we did the following: (1) added Paint and Textile categories to the other category; (2) calculated the percentage of material; and (3) calculated the average material composition from the two types, and used it instead of weight. The average material composition was tabulated in the study report (see Al-Quradaghi et al. 2020).

In implementing the model, the following assumptions were made: (1) the DF operates 10h per day (7:00 a.m. to 17:00 p.m.), (2) three Powerhand vehicle-recycling-system machines are used to dismantle the vehicles, and (3) average processing time for vehicles is 10 min.

Model Results

The connected network of exchange flows is simulated via the SIMIO software package (Joines and Roberts 2015). The produced results spanning one month (28 working days) are summarized as follows:

- Iron and Steel: 1,548
- Aluminum: 127
- Glass: 73
- Plastic: 207
- Tires: 56
- Battery: 25
- Fluids: 91
- Other: 157

As can be seen in the table, the DF processed 1,820 vehicles, and the material outcome from the network could be used for each corresponding industry as a resource. This is only an example of how one month of recycling ELVs produces a different quantity of materials that can be reused. If done for the long term, the EIP-4-ELV network serves in the reduction of raw material extractions and GHG emissions.

CONCLUSIONS

EIPs promote the shift from the traditional linear to the circular model, where by-products and waste can be reused. The EIP literature covers multidisciplinary areas, including optimization, lifecycle assessments, policy implementation, social networks, and typology. The International Framework for Eco-Industrial Parks provided by the UNIDO, GIZ, and WBG reports serves as a guide for decision makers on important components to achieve maximal benefits economically, environmentally, and socially. However, as indicated in the report, the International Framework for Eco-Industrial Parks provides only strategic details for the EIP requirements and does not translate them into an existing EIP.

On the other hand, the frameworks for EIPs that are provided in the literature are very comprehensive. Hence, there is increasing need for a straightforward framework. The proposed framework illustrates the foundations of designing EIPs. It shows general step-by-step actions to be discussed at the very early stages of designing EIPs. The framework answers fundamental questions that decision makers need to consider for developing and designing EIPs. In a simple, clear, and step-by-step strategy, the framework lays out necessary actors in the process of designing EIPs. The framework requires primary data that can easily be collected from each entity that forms EIPs.

The proposed framework bridges the gap in the literature and provides a generalized framework for the design of EIPs. The framework was employed to solve the ELV problem (EIP-4-ELVs). As a result of applying the framework in EIP-4-ELVs, a connected network of exchanges was built. The outcomes represent the amount of different types of materials. If applied to solve the ELV problem, the framework can create a connected network that produces different types of materials. By using EIP-4-ELVs, the network prevents using more natural sources and depends on some percentage of the by-product exchange from other industries.

REFERENCES

Aid, G., Eklund, M., Anderberg, S., and Baas, L. (2017), "Expanding Roles for the Swedish Waste Management Sector in Inter-Organizational Resource Management," *Resources, Conservation and Recycling*, Vol. 124, pp. 85–97.

Al-Quradaghi, Shimaa, Zheng, Quipeng P., and Eklamel, Ali (2020), "Generalized Framework for the Design of Eco-Industrial Parks: Case Study of End-of-Life Vehicles," *Sustainability*, Vol. 12, p. 6612, doi: 10.3390/su12166612.

Andersson, M. (2016), *Innovating Recycling of End-of-Life Cars*, Chalmers University of Technology, Göteborg, Sweden.

Andiappan, V., Tan, R. R. and Ng, D. K. S. (2016), "An optimization-based negotiation framework for energy systems in an eco-industrial park," *Journal of Cleaner Production*, Vol. 129, pp. 496–507.

ARA (2012), *Automotive Recycling Industry: Environmentally Friendly, Market Driven, and Sustainable*, ARA, Manassas, VA.

Baker, B. A., Brookside, N. J., Woodruff, K. L., Morrisville, P., Naporano, J. F., and Fells, N. J. E. (October 12, 1995), "Automobile Shredder Residue (ASR)

Separation and Recycling System," *Journal of Industrial Ecology*, U.S. Patent WO1995026826A1.

Behera, S. K., Kim, J.-H., Lee, S.-Y., Suh, S., and Park, H.-S. (2012), "Evolution of 'Designed' Industrial Symbiosis Networks in the Ulsan Eco-Industrial Park: 'Research and Development into Business' as the Enabling Framework," *Journal of Cleaner Production*, Vol. 29–30, pp. 103–112.

Boix, M., Montastruc, L., Azzaro-Pantel, C. and Domenech, S. (2015), "Optimization methods applied to the design of eco-industrial parks: A literature review," *Journal of Cleaner Production*, Vol. 87, pp. 303–317.

Boons, F., Spekkink, W., and Mouzakitis, Y. (2011), "The Dynamics of Industrial Symbiosis: A Proposal for a Conceptual Framework Based upon a Comprehensive Literature Review," *Journal of Cleaner Production*, Vol. 19, pp. 905–911.

Che, J., Yu, J., and Kevin, R. S. (2011), "End-of-Life Vehicle Recycling and International Cooperation between Japan, China and Korea: Present and Future Scenario Analysis," *Journal of Environmental Sciences*, Vol. 23, pp. S162–S166.

Chertow, M. R. (2000), "Industrial Symbiosis: Literature and Taxonomy," *Annual Review of Energy and the Environment*, Vol. 25, pp. 313–337.

Chertow, M. R. (2007), "Uncovering Industrial Symbiosis," *Journal of Industrial Ecology*, Vol. 11, p. 20.

Choi, J. K., Stuart, J. A., and Ramani, K. (2005). "Modeling of Automotive Recycling Planning in the United States," *International Journal of Automotive Technology*, Vol. 6, pp. 413–419.

Curlee, T. R., Das, S., Rizy, C. G., and Schexnayder, S. M. (1994), *Recent Trends in Automobile Recycling: An Energy and Economic Assessment*, Oak Ridge National Laboratory, Oak Ridge, TN.

D'Adamo, I., Gastaldi, M., and Rosa, P. (2020), "Recycling of End-of-Life Vehicles: Assessing Trends and Performances in Europe," *Technological Forecasting and Social Change*, Vol. 152, pp. 905–911.

de Almeida, S. T. and Borsato, M. (2019), "Assessing the Efficiency of End of Life Technology in Waste Treatment – a Bibliometric Literature Review," *Resources, Conservation and Recycling*, Vol. 140, pp. 189–208.

Dumoulin, F., Wassenaar, T., Avadi, A., and Paillat, J. (2016), "A Framework for Accurately Informing Facilitated Regional Industrial Symbioses on Environmental Consequences," *Journal of Industrial Ecology*, Vol. 21, pp. 1049–1067.

Edwards, C., Bhamra, T., and Rahimifard, S. (May 31–June 2, 2006) "A Design Framework for End-of-Life Vehicle Recovery." In Proceedings of the 13th CIRP International Conference on Life Cycle Engineering, Leuven, Belgium, pp. 365–370.

Ehrenfeld, J. and Gertler, N. (1997), "Industrial Ecology in Practice; The Evolution of Interdependence at Kalundborg," *Journal of Industrial Ecology*, Vol. 1, pp. 67–79.

Eilering, J. A. M. and Vermeulen, W. J. V. (2004), "Eco-Industrial Parks: Toward Industrial Symbiosis and Utility Sharing in Practice," *Progress in Industrial Ecology: an International Journal*, Vol. 1, pp. 245.

European Parliament and Council (2000), *Directive 2000/53/EC on End-of-Life Vehicles*, Official Journal of the European Communities European Parliament and Council: Brussels, Belgium, L269; pp. 34–42.

Felicio, M., Amaral, D., Esposto, K., and Durany, X.G. (2016), "Industrial Symbiosis Indicators to Manage Eco-Industrial Parks as Dynamic Systems," *Journal of Cleaner Production*, Vol. 118, pp. 54–64.

Frosch, R. A. (1994), "Industrial Ecology: Minimizing the Impact of Industrial Waste," *Physics Today*, Vol. 47, pp. 63–68.

Frosch, R. A. and Gallopoulos, N. E. (1989), "Strategies for Manufacturing," *Scientific American*, Vol. 261, pp. 144–153.

Garcia, J., Millet, D., and Tonnelier, P. (2015), "A Tool to Evaluate the Impacts of an Innovation on a Product's Recyclability Rate by Adopting a Modular Approach: Automotive Sector Application," *International Journal of Vehicle Design*, Vol. 67, pp. 178–204.

Gopinath, A., Bahurudeen, A., Appari, S., and Nanthagopalan, P. (2018), "A Circular Framework for the Valorisation of Sugar Industry Wastes: Review on the Industrial Symbiosis between Sugar, Construction and Energy Industries," *Journal of Cleaner Production*, Vol. 203, pp. 89–108.

Graedel, T. E., Allwood, J., Birat, J.-P., Reck, B. K., Sibley, S. F., Sonnemann, G., Buchert, M., and Hagelüken, C. (2011), *Recycling Rates of Metals – A Status Report, A Report of the Working Group on the Global Metal Flows to the International Resource Panel*, UNEP, Paris, France.

Gu, C., Leveneur, S., Estel, L., and Yassine, A. (2013), "Modeling and Optimization of Material/Energy Flow Exchanges in an Eco-Industrial Park," *Energy Procedia*, Vol. 36, pp. 243–252.

Haskins, C. (2007), "A Systems Engineering Framework for Eco-Industrial Park Formation," *Systems Engineering*, Vol. 10, pp. 83–97.

Heeres, R. R., Vermeulen, W. J. V., and de Walle, F. B. (2004), "Eco-Industrial Park Initiatives in the USA and the Netherlands: First Lessons," *Journal of Cleaner Production*, Vol. 12, pp. 985–995.

Jiao, W. and Boons, F. (2014), "Toward a Research Agenda for Policy Intervention and Facilitation to Enhance Industrial Symbiosis Based on a Comprehensive Literature Review," *Journal of Cleaner Production*, Vol. 67, pp. 14–25.

Joines, J. A. and Roberts, S. (2015), *Simulation Modeling with SIMIO: A Workbook*, 4th ed., SIMIO LLC, Sewickley, PA.

Kahane, C. J. (2003), *Vehicle Weight, Fatality Risk and Crash Compatibility of Model Year 1991–99 Passenger Cars and Light Trucks*, National Highway Traffic Safety Administration, Springfield, VA.

Karagoz, S., Aydin, N., and Simic, V. (2019), "End-of-Life Vehicle Management: A Comprehensive Review," *Journal of Material Cycles and Waste Management*, Vol. 22, pp. 416–442.

Konstantinova, Y., Johannes, E., and Vejrum, B. (2019), "Dare to Make Investments in Industrial Symbiosis? A Conceptual Framework and Research Agenda for Developing Trust," *Journal of Cleaner Production*, Vol. 223, pp. 989–997.

Kuznetsova, E., Zio, E., and Farel, R. (2016), "A Methodological Framework for Eco-Industrial Park Design and Optimization," *Journal of Cleaner Production*, Vol. 126, pp. 308–324.

Liu, C. and Côté, R. (2017), "A Framework for Integrating Ecosystem Services into China's Circular Economy: The Case of Eco-Industrial Parks," *Sustainability*, Vol. 9, pp. 1510.

Mat, N., Cerceau, J., Shi, L., Park, H. S., Junqua, G., and Lopez-Ferber, M. (2016), "Socio-Ecological Transitions toward Low-Carbon Port Cities: Trends, Changes and Adaptation Processes in Asia and Europe," *Journal of Cleaner Production*, Vol. 114, pp. 362–375.

Maudet, C., Yannou-Le Bris, G., and Froelich, D. (2012), "Integrating Plastic Recycling Industries into the Automotive Supply Chain," *HAL*, Vol. 13, pp. 71–89.

Meerkamm, H. (2007), "Design for X – a Core Area of Design Methodology," *Journal of Engineering Design*, Vol. 5, pp. 165–181.

Paul, R. (2009), "End-of-Life Management of Waste Automotive Materials and Efforts to Improve Sustainability in North America," *WIT Transactions on Ecology and the Environment*, Vol. 120, pp. 853–861.

Piaszczyk, C. (2011), "Model Based Systems Engineering with Department of Defense Architectural Framework," *Systems Engineering*, Vol. 14, pp. 305–326.

Reuter, M. A., Hudson, C., van Schaik, A., Heiskanen, K., Meskers, C., and Hagelüken, C. (2013), *Metal Recycling: Opportunities, Limits, Infrastructure, a Report of the Working Group on the Global Metal Flows to the International Resource Panel*, UNEP, Paris, France.

Romero, E. and Ruiz, M. C. (2013), "Framework for Applying a Complex Adaptive System Approach to Model the Operation of Eco-Industrial Parks," *Journal of Industrial Ecology*, Vol. 17, pp. 731–741.

Simic, V. (2013), "End-of-Life Vehicle Recycling – A Review of the State-of-the-Art," *Recikliranje vozila na Kraj. životnog ciklusa—Pregl. Najsuvremnijih Znan. Rad.*, Vol. 20, pp. 371–380.

Song, X., Geng, Y., Dong, H., and Chen, W. (2018), "Social Network Analysis on Industrial Symbiosis: A Case of Gujiao Eco-Industrial Park," *Journal of Cleaner Production*, Vol. 193, pp. 414–423.

Sopha, B. M., Fet, A. M., Keitsch, M. M., and Haskins, C. (2009), "Using Systems Engineering to Create a Framework for Evaluating Industrial Symbiosis Options," *Systems Engineering*, Vol. 13, pp. 149–160.

Susur, E., Hidalgo, A., and Chiaroni, D. (2019), "A Strategic Niche Management Perspective on Transitions to Eco-Industrial Park Development: A Systematic Review of Case Studies," *Resources, Conservation and Recycling*, Vol. 140, pp. 338–359.

Tao, Y., Evans, S., Wen, Z., and Ma, M. (2019), "The Influence of Policy on Industrial Symbiosis from the Firm's Perspective: A Framework," *Journal of Cleaner Production*, Vol. 213, pp. 1172–1187.

Tessitore, S., Daddi, T., and Iraldo, F. (2015), "Eco-Industrial Parks Development and Integrated Management Challenges: Findings from Italy," *Sustainability*, Vol. 7, pp. 10036–10051.

Tian, J., Liu, W., Lai, B., Li, X., and Chen, L. (2014), "Study of the Performance of Eco-Industrial Park Development in China," *Journal of Cleaner Production*, Vol. 64, pp. 486–494.

Tonnelier, P., Millet, D., Richir, S., and Lecoq, M. (2007), "Is It Possible to Evaluate the Recovery Potential Earlier in the Design Process? Proposal of a Qualitative Evaluation Tool," *Journal of Engineering Design*, Vol. 16, pp. 297–309.

UNIDO, GIZ (2017), "WBG." In *An International Framework for Eco-Industrial Parks*, UNIDO, Danver, MA.

van Schaik, A. and Reuter, M. A. (2004), "The Optimization of End-of-Life Vehicle Recycling in the European Union," *JOM*, Vol. 56, pp. 39–43.

Vezzoli, C., Ceschin, F., Osanjo, L., M'Rithaa, M. K., Moalosi, R., Nakazibwe, V., and Diehl, J. C. (2018), *Designing Sustainable Energy for All. Sustainable Product-Service System Design Applied to Distributed Renewable Energy*, Springer, Cham, Switzerland, p. 230.

Villalba, G., Segarra, M., Chimenos, J. M., and Espiell, F. (2004), "Using the Recyclability Index of Materials as a Tool for Design for Disassembly," *Ecological Economics*, Vol. 50, pp. 195–200.

Watson, B. and Radcliffe, D. (2010), "Structuring Design for X Tool Use for Improved Utilization," *Journal of Engineering Design*, Vol. 9, pp. 211–223.

Yedla, S. and Park, H. (2017), "Eco-Industrial Networking for Sustainable Development: Review of Issues and Development Strategies," *Clean Technologies and Environmental Policy*, Vol. 19, pp. 391–402.

Yu, J., Wang, S., Toshiki, K., Serrona, K. R. B., Fan, G., and Erdenedalai, B. (2017), "Latest Trends and New Challenges in End-of-life Vehicle Recycling." In *Environmental Impacts of Road Vehicles: Past, Present and Future*, The Royal Society of Chemistry: London, UK, Vol. 44, pp. 174–213.

Zhang, Y., Duan, S., Li, J., Shao, S., Wang, W., and Zhang, S. (2017), "Life Cycle Assessment of Industrial Symbiosis in Songmudao Chemical Industrial Park, Dalian, China," *Journal of Cleaner Production*, Vol. 158, pp. 192–199.

Index